公共建筑与空间营造
Public Buildings and Placemaking

[丹] BIG建筑事务所等 | 编

于风军 林英玉 辛敏裕 罗茜 张可新 吴欣霓 周美含 | 译

大连理工大学出版社

004　建筑行业中的女性都去哪儿了？_ Martha Thorne

塑造我们在城市中的居住形式

008　塑造我们在城市中的居住形式 _ Anna Roos

014　YCONE住宅楼 _ Ateliers Jean Nouvel

030　L'Arbre Blanc公寓 _ Sou Fujimoto Architects + Nicolas Laisné + Dimitri Roussel + OXO Architectes

042　La Borda合作住房 _ Lacol

公共建筑与空间营造

056　公共建筑与空间营造 _ Angelos Psilopoulos

062　绿色广场图书馆和广场 _ Studio Hollenstein

074　de Maio街24号SESC综合体 _ Paulo Mendes da Rocha + MMBB Arquitetos

092　MÉCA文化中心 _ BIG

110　比奥比奥地区剧院 _ Smiljan Radic + Eduardo Castillo + Gabriela Medrano

128　高雄艺术中心 _ Mecanoo architecten

加建的不仅仅是空间

148　加建的不仅仅是空间 _ Phil Roberts

154　多伦多大学的丹尼尔斯大楼 _ NADAAA

174　自然生物多样性中心 _ Neutelings Riedijk Architects

192　AGORA癌症研究中心 _ Behnisch Architekten

206　CSN总部办公楼扩建项目 _ BGLA + NEUF Architectes

218　日内瓦鲁道夫·斯坦纳学校 _ LOCALARCHITECTURE

230　建筑师索引

004   Where are all the women?_Martha Thorne

Shaping the Way we Live in Cities

008   Shaping the Way we Live in Cities_Anna Roos

014   YCONE Residential Tower_Ateliers Jean Nouvel

030   L'Arbre Blanc_Sou Fujimoto Architects + Nicolas Laisné + Dimitri Roussel + OXO Architectes

042   La Borda Cooperative Housing_Lacol

Public Buildings and Placemaking

056   Public Buildings and Placemaking_Angelos Psilopoulos

062   Green Square Library and Plaza_Studio Hollenstein

074   SESC 24 de Maio_Paulo Mendes da Rocha + MMBB Arquitetos

092   MÉCA Cultural Center_BIG

110   Biobío Regional Theater_Smiljan Radic + Eduardo Castillo + Gabriela Medrano

128   Kaohsiung Center for the Arts_Mecanoo architecten

Additions That Add More Than Space

148   Additions That Add More Than Space_Phil Roberts

154   University of Toronto Daniels Building_NADAAA

174   Naturalis Biodiversity Center_Neutelings Riedijk Architects

192   AGORA Cancer Research Center_Behnisch Architekten

206   Expansion of the CSN Headquarters_BGLA + NEUF Architectes

218   Rudolf Steiner School of Geneva_LOCALARCHITECTURE

230   Index

建筑行业中的女性都去哪儿了?
Where are all the women?

Martha Thorne

我们这些在建筑和设计领域工作的人喜欢使用短句来帮助外行人理解我们所做的事情。我和其他人经常使用的最喜欢的一个短句是"问题比答案更重要"。当然,这意味着建筑师和设计师并非只是接受给定的参数寻求一维的解决方案,他们会根据项目的总体目标,从项目概要到具体要求的一切进行质疑。我经常说,这就是我们的特别之处,因为我们有能力去想象别人可能看不到的未来的可能性。我们不只是分析现在,我们还可以创造性地预想到未来。

为什么建筑领域的知名女性寥若晨星?这是我们应该问的问题吗?也许不是。纵观历史,一些属于20世纪的名字会立刻浮现在我们的脑海中,如艾琳·格雷、莉莉·赖希、简·雅各布斯、马里恩·马奥尼·格里芬或莉娜·博·巴迪。但除了这些人,还有谁能让我们了然于心、脱口而出呢?更重要的是,除了名字(或者在最糟糕的情况下是她的伙伴或丈夫的名字),我们还能记得什么呢?

近年来,普利兹克奖得主扎哈·哈迪德和妹岛和世等女性名字被作为一些女性取得巨大成功的例子。然而,这两位女性都被限定在一个标准模式下,她们都被打上了具有强烈的个性、没有一个传统的家庭生活、比较独立等等标签。我可以列举出无数来自世界各地的有才华的女性,她们正以诸多不同的方式从事着建筑行业,但她们没有时间或兴趣去宣传自己,使自己成为所谓的"名人",或者各种各样的媒体也许认为,她们所做的工作或她们所表现出的

Those of us who work in the fields of architecture and design like to repeat phrases that help laypeople understand what we do. One favorite phrase, that I and others use quite often, is "the question is more important than the answer". The meaning of this, of course, is that architects and designers don't merely accept the given parameters and seek a one-dimensional solution, but they question everything from the brief to the constraints to the overall goals of a project. I have often argued that this is what makes us special, as we have the ability to imagine future possibilities that others may not see. We do not just analyze the present, but we can creatively envision a future.

Why are there so few well-known women in the architecture field? Is this the question we should be asking? Perhaps not. When we look throughout history, there are some names from the 20th century that come immediately to mind, such as Eileen Gray, Lily Reich, Jane Jacobs, Marion Mahoney Griffin, or Lina Bo Bardi. But beyond this shortlist, who can we easily recall? And more importantly, what do we remember beyond a name (or in the worst case the name of her partner or husband)?

More recently women's names such as Zaha Hadid or Kazuya Sejima, both winners of the Pritzker Prize, are brought forward as examples that some women can be very successful. However, both of these figures are somehow within a standard format, characterized by a strong personality, without a traditional family with children, independent, etc… I can name numerous examples of talented women from around the world, currently practicing architecture, in many different ways, but either

价值不值得报道，不值得成为"今日头条"。你听说过Odile Decq、Carla Juaçaba、Carol Ross Barney、Marion Weiss、Farshid Moussavi、Kate Otten或Inês Lobo吗？她们只是众多建筑行业中充满才华的女性中的几个。

为了清楚地知道我们将何去何从，我想提出几个新的问题。这些问题包括：女性在建筑和设计领域的障碍是什么？今天，如果超过50%的建筑学学生是女性（其实这已经成为事实超过十年了），为什么她们在这个行业中没有得到充分的代表？为什么她们不能得到同工同酬的待遇呢？我们又可以采取什么行动来发掘职业女性的才能，并从她们所具备的知识和技能中得到裨益呢？

的确，女性进入这一行业的时间比男性晚，第一批建筑学专业的女性毕业生是在19世纪末20世纪初获得学位的。这些早期的女性毕业生显然是个例外，而不是常态。如果今天我们的学生中有一半或一半以上是女性，那么实现平等不就是时间的问题了吗？可惜的是，并不是这样。我认为，有几个因素仍然对女性在这一领域形成阻碍。我们的学校和专业决策机构的结构仍然偏向于过去的模式，而不是着眼于未来。固有的模式阻碍了建筑师或者设计师的变革，就像一个流行的神话：建筑师是一个寂寞的项目创造者，或者是从黑盒子里掷出设计方案的天才。

如果在此事实基础上我们再考虑一点，即与建筑相关的行业，如银行、建筑公司、投资者和开发公司，都是男性

they do not have time or interest to promote themselves as "celebrities" or the work they do and the values they espouse fall outside what the press considers worthy of headlines. Have you heard of Odile Decq, Carla Juaçaba, Carol Ross Barney, Marion Weiss, Farshid Moussavi, Kate Otten or Inês Lobo, just to name a few?

In an attempt to understand where we have come from and where we might be going, I prefer to frame new questions. These include such queries as, what are the barriers to women in architecture and design? Today, if more than 50% of architecture students are women (and have been for more than a decade), why aren't they fully represented in the profession? Why don't they receive equal salaries for equal work? And finally, what actions can we take to recognize talent and benefit from the knowledge and skills that women professionals bring to the table?

It is true that women came to the profession later than men, and the first women graduates in architecture received their diplomas in the late 19th and early 20th century. These early women graduates were clearly the exception, not the norm. If today half or more of our students are women, isn't it just a question of time to attain equality? Unfortunately, no. I believe that several factors still constitute barriers to women in the field. The structure of the decision-making bodies in our schools and professions is still skewed towards the past, rather than the future. The inertia of past stereotypes of what it takes to be an architect or designer is resistant to change, as is the popular myth that an

占主导地位的行业,那么女性很难涉足其中并获得应有的地位就不足为奇了。人们普遍认为,建筑师这个职业需要长时间地工作、召开没完没了的研讨会议、承受各种最后期限的压力,这常常迫使建筑师在没有固定工作时间表的情况下还要在周末工作。这让那些有家庭或希望在职业之外有私人生活的专业人士对建筑师这一行业望而却步。而这尤其给女性和母亲带来了不公平的负担。

虽然建筑学院试图在教学岗位上实现男女平等(但其结果往往不尽如人意),但他们仍然没有改变"师徒"的传统教学方法,也没有在课程设置、课程内容或学生使用的书目中加入新的女性声音。例如,全是男性的特邀讲师名单,或全是男性的评审团成员,这样令人习以为常的"小事"往往会强化旧有模式,在不知不觉中限制了女性。

最后,建筑行业正处在一个超越性别的十字路口。新技术、即时通信和全球化意味着我们不再局限于以传统方式共享信息和知识。我们可以通过互联网的多种媒体来交流和表达我们的想法。这也意味着,我们的工作方式不受地域限制,我们能够跨越距离,以新的方式、新的角色与他人合作。建筑信息模型(BIM)这个小小的例子就能说明这一点。建筑信息模型(BIM)可以共享设计和施工过程中的每一步信息,这就限制了建筑师通过简单控制信息而获得的权利。

architect is a lone creator of projects or a genius who pulls designs out of a black box.

If we add to this the fact that the industries around architecture, such as banks, construction companies, investors and development companies are heavily male-dominated, it is no surprise that women have a difficult time entering and taking their rightful places. The accepted belief that the profession requires long hours, charrettes and deadlines that often force architects to work without a set schedule and on weekends, limits those professionals who have families or want a private life in addition to their career. And this has placed an unfair burden on women and mothers.

Schools of architecture, while trying to achieve parity in teaching positions (not always successfully), still do not revamp the traditional teaching method of "master and disciple" nor do they require new, female voices in course content, curricula or bibliographies used by students. Small things such as a roster of male special guest lecturers or all-male juries reinforce old models and inadvertently limit women.

Finally, the architecture profession is at a crossroads that goes way beyond gender. New technologies, instantaneous communication, and globalization mean that we are no longer limited to information and knowledge shared in traditional ways. We can communicate and make our ideas known through multiple media via the internet. It also means that the way we work is not bounded by geography, but we are able to collaborate across distances, in new ways, and with new roles. One small example is evidenced by building information modeling (BIM), which, by sharing information at every step of the

现在，在建筑行业中灌输新的价值观和新的工作方式变得至关重要。应该使协作、团队努力以及团队里除了设计师之外的其他角色同样得到制度化奖励。我们应该意识到，在建筑设计领域，加班加点长时间工作，有时候还有没日没夜"百米冲刺"这样的情况，会令那些希望工作之余拥有个人生活的人，尤其是那些有家庭的人望而生畏，这也无益于贡献出最佳设计或解决方案。

我们不仅要在建筑设计事务所内实现各种改变，在既定的奖项体系中，同样需要改变用来评判同事和其他建筑师的标准。所幸的是，著名建筑师的时代已经结束了。现在是这个行业在寻求性别平等和管理机制多样性方面发挥领导作用的时候了，不仅因为这是正确的做法，而且因为这是明智的举措。去改变旧有价值观念，去提倡并崇尚协作、包容性设计，在设计过程和设计产品中体现可持续性等我们所拥护的价值观，自然会带来更加多样化和人性化的做法。它还将传递这样一个信息：优秀的设计是许多人努力和严格执行的结果，他们发出不同的声音，接受不同的任务，但所有这些都是必要的，只有同心协力才能达成。最后，这将意味着我们不会白白地失去许多学了多年建筑和相关学科的人的才能，不会埋没他们的才华，不会使他们感觉在专业领域之内无用武之地。我们认为，在21世纪，我们面临着巨大且至关重要的全球性挑战，要应对这些挑战，不能没有女性的思想、投入和努力。显而易见，建筑领域的新秩序将福泽我们所有的人。

design and construction processes, is limiting the power of the architect that "he" had by simply controlling information.

It now becomes crucial to instill new values and new ways of working within the practice of architecture. Collaboration, team efforts and rewarding those in roles beyond the design architect should be instituted. Realizing that working long hours in a charrette like "sprint" does not favor those who have a life outside the office, especially those with families, nor does it contribute to the best designs or solutions.

As within the office, also in the system of established prizes and awards, we need to shift the criteria we use to judge colleagues and other architects. Thankfully, the era of the starchitect has ended. Now is the time for the profession to lead in terms of seeking gender equality and embracing diversity, not only because it is the right thing to do, but because it is the smart thing to do. Shifting the values we espouse to include and even highlight such things as collaboration, inclusive design, and sustainability in our processes and outcomes will naturally lead to more diverse and humane practices. It will also send a message that good architecture is the result of rigorous processes and efforts by many, with different voices, undertaking different tasks, but all necessary and only possible when working together. Finally, it will mean that we will not uselessly lose the talent of so many people who have studied architecture and related disciplines for many years only to be left outside the profession or hidden in the shadows. The global challenges facing us in the 21st century are too great and too important to think that we can tackle them without the ideas, input, and efforts of women. Clearly, a new order in architecture will benefit all of us.

塑造我们在城市中的居住形式

Shaping We Live

"家",这样一个小小的词,充满了意义:是住所,是身份,是家庭,是幸福,是财产,是避难所,是庇护所等。无论是在生理上还是在心理上,"家"都传达着人们对生活的渴望、寄托、归属与希望……建筑师们肩负着为广大民众设计家园的重大责任,但是如何才能在设计中体现出那些令人产生共鸣的理念呢?现在,住房问题仍然是一个十分紧迫的问题。随着移民源源不断地涌入市中心,全球都面临着越来越大的压力——需要以经济实惠的价格为移民提供优质的住宿条件。住房问题不仅仅是一个建筑难题,而且是一个严肃的问题,涉及城市规划、社会和政治。进入21世纪,在一些富裕城市(像伦敦、巴黎、洛杉矶等)中,

For such a small word, home is laden with meaning: dwelling, identity, family, wellbeing, property, refuge, shelter, both physically and psychologically, aspiration, rootedness, belonging, hope… How can architects embody all these evocative ideas in their designs without being overwhelmed by the enormity of the responsibility of designing homes for the masses? Now housing remains a pressing issue. With the non-stop stream of migrants and immigrants to urban centers worldwide grows the pressure to provide accommodation with high quality at affordable prices. Housing is not only an architectural conundrum, but a serious urban planning, social and political issue. It is shocking in the twenty-first century to see people sleeping rough in shop

YCONE住宅楼_YCONE Residential Tower / Ateliers Jean Nouvel
L'Arbre Blanc公寓_L'Arbre Blanc / Sou Fujimoto Architects + Nicolas Laisné + Dimitri Roussel + OXO Architectes
La Borda合作住房_La Borda Cooperative Housing / Lacol
塑造我们在城市中的居住形式_Shaping the Way We Live in Cities / Anna Roos

the Way in Cities

仍有人在商店门口或桥下露宿是一件令人十分震惊的事。如果住房权利是一项人权的话，那么我们现在应该做的就是大批量地设计、建造房屋。对于市中心的低收入家庭而言，他们能支付得起的住宅应该位于公共设施齐全、交通便利的市中心，而不是周边的贫民窟。但是，由于空地资源短缺，市中心的土地价格是很高的，因此，为了在资金上具有可行性，一个项目应该以高密度建设为设计方案。那么在21世纪，经济实惠的高密度居住形式是怎样的呢？本书选择了最近在欧洲建造的三个示范性项目，以展示存在于市中心的那些有趣的、甚至富有诗意的居住形式。

doorways and under bridges in wealthy cities, like London, Paris and Los Angeles. If housing is a human right, and it is, then we need, right now, to design houses en masse. Houses that low income families in urban centers can afford, not in ghettoes on the periphery, but centrally, close to amenities and public transport. A central location though, implies high land values due to the lack of open land, thus in order to make a project financially viable, it ought to be of high-density. What might affordable high-density living look like in the twenty-first century? This book has chosen three exemplary projects recently built in Europe to showcase what interesting, even poetic, approaches exist to inner city living.

塑造我们在城市中的居住形式
Shaping the Way We Live in Cities

Anna Roos

"这是一座软性城市；它在等待着新的身份印记。无论是好还是坏，它都会邀请您来重新改造自己，以塑造出一个您适宜居住的环境。"——乔纳森·拉班，《软性城市》，1974年。

解决城市高质量廉租房问题是21世纪最紧迫的社会问题之一。我们的建筑结构涉及的不仅是关于功能的问题，其中还交织着社区生活、社会互动和各种关系，它们共同构成了城市社区的蓝图。现如今，建筑师们面临着这样的挑战——以经济实惠的价格建造出更多的住房。然而这绝非易事，因为在城市中高昂的土地价格会自动抬高房价，甚至土地价格会常常超过建筑成本。同时，住房需求常常超过供应，这也会抬高房价。投资高寄希望于通过住房项目获得更多收益，因此也会抬高房价。似乎低利率可以降低建筑成本，但当利率较低时，更多的人会投资房地产，从而增加需求、抬高房价。这些都是建筑师无法改变的因素。只有规划者与政治家才有能力推动变革，帮助人们应对住房问题所面临的巨大变化。

建筑师们已经不再把住宅看作纯粹的技术性"居住机器"。住宅应该是高效的且耐用的，但是它们还应该为我们带来美好的情感体验与幸福感。可以这样认为，在20世纪，柯布西耶可能是对住房有最大影响的建筑师。对于住房，他常常有一些比较激进的看法。他认为家是改革社会的关键。柯布西耶认为，房子就像一台机器一样，是集实用性、高效性与美观性于一体的，如同"一台耐用的打字机"。但是随着时间的推移，柯布西耶的设想被人们指责为非人性化的且毫无灵魂的，是充满破坏公物行为与犯罪的贫民窟。现如今，建筑师已经意识到，他们的设计项目不仅需要满足技术上的要求，还需要使住房具有感召力，令人们在情感上得到满足。

社会和人口的变化表明，我们需要设计灵活的空间。理想情况下，住房不应太过规范化，而应该更具选择性，以便居住在那里的人能够进行调节与适应。仔细研究过日本的建筑后发现，他们的建筑大多通过减少功能性来增加灵活性，以实现更大的自由，增加更多的可能性。毕竟一个空间不需要全天24小时发挥一个功能。避免将空间高度专业化，以及将结构和内部装修分开，

"… the city goes soft; it awaits the imprint of an identity. For better or worse, it invites you to remake it, to consolidate it into a shape you can live in." – Johnathan Raban, Soft City, 1974.

Affordable housing with good quality in cities is one of the most pressing social questions of this century. Our built fabric is not only a question of function, but within it are woven concepts of communal living, social interactions and relationships that create the tapestry of urban communities. Architects face the challenge of producing masses of housing at affordable prices. This is no mean feat, as the high price of land in cities automatically elevates housing prices. Land prices are often even greater than the cost of construction. As demand often outstrips supply, this also inflates prices. Investors pin their hopes on profiting from housing projects, thus also pushing house prices upward. Low interest rates might appear to reduce building costs, but when interest rates are low, more people look to invest in property, thus increasing demand and house prices. These are factors that architects cannot influence. It is planners and politicians who have the power to push for change and help to tackle the seismic changes that face housing.

Architects have moved away from the idea of dwelling being purely technical "machine for living in". Houses ought to be efficient and serviceable, but they also need to appeal to our emotions and sense of beauty and wellbeing. Corbusier, the architect who arguably had the greatest influence on housing in the twentieth century, had some radical ideas on housing; he saw the home as a key to reforming society. Corbusier envisaged the house as practical, efficient and as beautiful as a machine, as "serviceable as a typewriter". But over time Corbusier-inspired schemes came to be reviled as inhuman and soulless, ghettoes of vandalism and crime. Today architects acknowledge the need for their projects to do more than merely fulfill technical requirements, our homes have to appeal to our senses and make us feel emotionally contented.

Social and demographic changes imply the design of flexible spaces. Ideally, housing should not be too prescriptive, but suggest or propose and include an element of chance so that appropriation and adaptation

可以使空间随着时间的推移而改变功能。这些多功能空间具有改变功能的能力，因此它们具有更高的价值。对于适应经济、社会与技术发展带来的变化而言，这样的灵活性以及去规范化是至关重要的。

　　当代住房的设计中，一个十分吸引人的例子便是由让·努维尔建筑事务所最新设计的位于里昂的住宅项目——一个关于18层住宅楼的设想（14页）。精美的白钢结构和半透明与透明的玻璃使建筑物显得虚无缥缈。努维尔非常谨慎地将高层建筑嵌入了原有建筑物的整体中。该建筑外立面那淡淡的色彩使其与周围的城市建筑（附近大多数的建筑物都是白色的）融在了一起。此建筑被称为YCONE住宅楼，位于索恩河和罗纳河的交汇处，是拉康弗伦斯区更新改造总体规划的一部分。建筑内部的混合使用空间（包括办公区）鼓励人们开展一些商业活动，一层空间向公众开放，一层还有一个露台花园。该建筑及开放式公共花园中的各种功能令这里充满了生机与活力，可以举办各种活动。努维尔将这座建筑视为"文化温床"，在这里，"许多不同的个人和社会行动者相遇了"。

　　建筑上层的立面像书页一样逐渐收起、消失，打造出脆弱的效果，并在"固有的"立面和阳台"表皮"之间创造了一个受保护的阳台空间，这是日本人可能称之为"ma"的"阴性"区域。建筑上各种色彩交织在一起——柔和的粉红色、奶油般的黄色和凉爽的蓝色与半透明与透明的玻璃共同创造出了一座有趣的建筑，一座三维的抽象雕塑，赋予了该建筑一个独特的身份。如果建筑物是"穿着"它们的外立面，那么YCONE一定是身披精美的三维高级定制礼服，在城市景观中作为一个美丽的形象，吸引着人们的目光。

　　众所周知，建筑业是向大气中排放二氧化碳的主要元凶之一。为了减少人类的碳足迹，同时控制成本，世界各地的城市正在探索新的紧凑型生活方式。合作住房以经济实惠的价格为人们提供了紧凑型的社区生活，侧重于社区型生活形式而非私人方式。与任何建筑策略相比，拥有更多共享公共空间而非私人空间可以在更大程度上降低建筑成本与碳排放。共享空间的另一个好处就是可以避免孤独和孤立，增加人们的幸福感。

are enabled by the people who live there. Taking a careful look at Japanese architecture, where flexibility is built in through absence of function can enable the freedom of many possibilities. After all a space does not need to be dedicated to a single function over a 24-hour period. Avoiding highly specialized spaces and separating structure and interior fitout allows spaces to change function over time. These multi-functional spaces have a higher value due to their ability to change function. This flexibility and lack of prescription is essential to be able to accommodate change: economic, social and technical.

A fascinating example of contemporary housing is Ateliers Jean Nouvel's latest housing project in Lyon, which is an ephemeral eighteen-story vision (p.14). Layers of delicate white steel structure and translucent and transparent glass dematerialize the building. With great care, Nouvel has inserted the high rise into the ensemble of preexisting buildings. The lightness of color ties the building into its surrounding urban fabric where most of the neighboring buildings are white. YCONE, as the building is called, forms part of an urban renewal master plan of La Confluence, where the Saône and Rhône rivers converge. Mix-used spaces, including office areas, encourage commercial activity and open the building to the public sphere on the ground level, where there is a terraced garden. This variety of functions housed in the building and the open public gardens enlivens the precinct with human activity. Nouvel sees the building as a "cultural hotbed" where "many different individuals and social actors meet".

The facade on the upper stories peals away, like the pages of a book, emphasizing the effect of fragility and creating a partly protected balcony space between the facade "proper" and the balcony "skin", an in-between, "negative" area that the Japanese might refer to as "ma". Daubs of color, soft pinks, buttery yellows and cool blues and panels of translucent and transparent glass create an intriguing three-dimensional abstract sculpture of the building, giving it a unique identity. If buildings are "dressed" in their facades, then YCONE is clad in a sublime three-dimensional haute couture robe, drawing attention to itself as a figure of beauty within the cityscape.

It is common knowledge that the building industry is one of the main culprits in emitting CO_2 gas into the atmosphere. In order to reduce our carbon footprint, while also reining in costs, new forms of compact living are

从根本上来说，重新考虑我们的生活方式会影响建筑方案的制定和设计。通过认真地考虑设计方案，用户很可能会同意其他解决方案，并放弃不必要的奢侈品以节省住房成本。例如，在欧洲，许多新的住房计划都禁止使用私家汽车。如果你放弃拥有私家汽车，转而使用公共交通工具，如开共享汽车或骑自行车，便意味着你可以放弃昂贵的地下车库，从而降低住房成本。此外，还有一些激进的例子，比如日本的F.O.B.A.项目，紧凑型公寓甚至没有任何厨房或浴室，居民使用餐厅和其他公共设施来满足相应需求。提供经济实惠的公寓的合作住房计划正在许多欧洲城市取得进展。

La Borda合作住房（42页）是Lacol与用户密切合作设计的合作住房计划。La Borda位于巴塞罗那的一个老工业区，是一块向公众开放的公有土地，目前正在进行升级、改造。

事实上，该项目由居民带头，从计划的构思到实现，居民都深入地参与其中。这意味着在建筑完工之前，人们就已经形成了一个互动的社区。用户的参与意味着将来的居民可以从一开始就向建筑师表达他们的特定需求。28套两室、三室和四室的公寓单元围绕一个光线明亮的中央庭院布置。庭院的设计灵感来自西班牙中部和南部传统的"corralas"式住房类型，是一种可以让人们互动和放松的住房形式。居民拥有公共空间这样的额外优势，可以弥补减少私人空间的"损失"。这种共享式的生活包括厨房/餐厅、多功能空间、客房、储藏室、洗衣房、工作空间和屋顶露台。他们的目标之一便是在施工阶段以及竣工后，在建筑物的使用寿命内，尽可能地减少对环境的影响。该设计采用被动式策略，旨在以最低的能耗达到较高的供暖水平，同时也使居民拥有更低的用房支出。

通过将木材、半透明波纹板和轻钢结构结合使用，Lacol创造出了一个多层内部庭院和交通流线空间，形成了合作住房计划的核心。庭院设有弧形玻璃屋顶用来防雨，屋顶上还设有开放式玻璃片以供通风，因此室内阳光充足，也不会过热。木材的广泛使用营造了一种温馨、宜人的氛围，居民可以用色彩缤纷的盆栽植物和时髦的家具，创造出属于自己的个性化空间。La Borda合

being explored in cities around the world. Cooperative housing offers compact, communal living at affordable prices, and focuses on community rather than private ownership. Sharing communal rather than owning private space can reduce building costs and carbon emissions to a greater degree than any construction strategy. Another advantage of sharing space is that loneliness and isolation are avoided and feelings of wellbeing are enhanced. Fundamentally rethinking the way we live impacts the brief and the design. By considering the brief critically, users might well agree to alternative solutions and forgo unnecessary luxuries to save costs. For instance, car ownership in many new housing schemes in Europe is forbidden. If you do away with car ownership in favor of public transport, car sharing or cycling means you can forgo expensive subterranean garages thereby reducing costs. There are radical examples like F.O.B.A. in Japan where the compact units do not even have any kitchens or bathrooms, residents using restaurants and public amenities instead. Cooperative housing schemes that offer affordable apartments are gaining ground in many European cities.

Designed in close collaboration with the users, Lacol's cooperative housing scheme, La Borda (p.42) is wedged in an open, publicly owned allotment in an old industrial area in Barcelona which is being upgraded and renewed. The fact that the residents spearheaded the project and were deeply involved in the schemes from its conception through to its realization meant that a community was formed even before the building was completed. This user participation meant that future residents could voice their specific needs to the architects from the onset. Twenty-eight, two-, three- and four-room units are arranged around a light-filled central courtyard inspired by the traditional "corralas" typology of housing in central and southern Spain, a place of interaction and relaxation. To mitigate the "loss" of private space, residents have the added advantage of communal space. This shared living includes kitchen/dining room, multipurpose space, guest rooms, storage, laundry, working spaces, and a roof top terrace. One of their objectives was to build with the lowest environmental impact possible, both during the construction phase, but also once completed, during the lifespan of the building. The design achieves high levels of thermal comfort with the minimum of energy consumption using passive strategies, in order also to make the units more affordable.

By combining timber with translucent corrugation and a lightweight steel structure, Lacol has created a multi-storied interior courtyard and circulation space that form the heart of the cooperative housing scheme. The

作住房是体现共享空间优势的有力证据,充分说明了在减少碳足迹的同时,建筑可以帮助人们创建繁荣的城市社区。

位于法国地中海沿岸蒙彼利埃的L'Arbre Blanc公寓(30页)采用的设计是建筑设计竞赛中的获胜方案。人们认为这一建筑设计可以丰富该城市的建筑遗产。与努维尔设计的住宅楼一样,这座十七层的建筑也是白色的,它的设计灵感来源于大自然。OXO建筑师事务所团队与日本建筑师藤本壮介在设计中使用了比喻的手法——将建筑理念与树木联系起来。阳台就像从树干上延伸出来的枝条,而遮阳板则像巨大的叶子一样"从立面上生长出来并保护着立面"。长达7.5m的悬臂式阳台给人一种眩晕的感觉,将室内生活空间向外延伸,人们站在上面可以俯瞰城市的景观。同时这些阳台还遮住了外立面,形成了一种"保护性面纱"。这对于一年中有80%的时间可以沐浴在阳光下的环境来说是一种实用的解决方案。阳台上方的悬垂板条为人们提供了阴凉的环境。无数的凸出构件分解了风力的作用,有利于空气"更加和谐"地循环。

为了向公众开放该建筑,一层建了一间美术馆;此外,屋顶酒吧可以为人们提供城市全景观赏平台,居民和公众皆可在此欣赏无限风光。邻近的莱兹河河畔建有一座景观公园,进一步扩展了公共空间。建筑师希望"通过让人们对住宅楼拥有实际所有权,使这座建筑成为蒙彼利埃人民的骄傲,也成为一个旅游景点"。

YCONE住宅楼、La Borda合作住房和L'Arbre Blanc公寓这三个项目以各自的方式说明了建筑如何在高密度城市地区实现新的高密度居住形式。这些建筑可以发挥多种功能——居住、零售和社交,展示了高质量的多功能住房如何帮助城市里的工业地区重获新生。诸如La Borda合作住房之类的方案展示了居民的需求如何能够超越纯粹的利润动机,而YCONE住宅楼和L'Arbre Blanc公寓则说明了创新设计如何能够创造强烈的认同感,帮助城市中心的老工业区重新焕发生机与活力。这些高层住房设计是极好的例子,说明了建筑师如何创建生机勃勃的社区和经济适用房,如何在住宅建筑中使用智能、可持续的创新设计,而蓬勃壮大的人口在这样的住房中居住都会感到骄傲和快乐。

courtyard is shielded from rain by a curved glazed roof with open glass fins for ventilation, thus allowing in ample daylight without overheating it. The extensive use of timber creates a warm, congenial atmosphere, where residents are able to create their own individualized entries with colorful pot plants and funky furniture. La Borda is concrete evidence of the advantages of sharing space and illustrates how architecture can help to create thriving urban communities while at the same time reducing our carbon footprint.

Built from a competition winning design, L'Arbre Blanc in Montpellier (p.30) on France's Mediterranean coast was envisaged as an enrichment of the city's architectural heritage. Like Nouvel's tower, this seventeen-story building is also an essay in white, a vision to behold inspired by nature. The team at OXO Architects together with Japanese architect, Sou Fujimoto used the metaphor of a tree for their design with balconies that branch off the trunk, and shades that "sprout out of and protect its facade" like enormous leaves. Deep, 7.5 meter-cantilevered balconies create a vertiginous feeling, while affording spectacular views over the cityscape and extending the interior living space outside. They also shade the facades, creating a kind of "protective veil", a practical solution for an environment that is bathed in sunshine for 80 percent of the year. Slatted overhangs above the balconies afford shade. The myriad of projections break up the wind and help air to circulate "more harmoniously".

In order to open the building to the public, there is an art gallery on the ground floor and rooftop bar with panoramic views of the city, which can be enjoyed by residents and the public alike. A landscaped park along the adjoining Lez River extends the public domain. The architects hope that "by allowing people to take physical ownership of the tower, it will become an object of pride for the people of Montpellier and a tourist attraction."

Each in their own way, the three projects, YCONE, La Borda and L'Arbre Blanc, illustrate how architecture can shape new ways of high density living in urban areas with high density. They show how high quality, mixed-use housing can help to regenerate industrial, inner city areas, allowing multiple functions: domestic, retail and social. Schemes like La Borda show how the needs of tenants can override the pure motivation for profit, while YCONE and L'Arbre Blanc illustrate how innovative design can create a strong identity and help regenerate industrial areas in urban centers. These high rise housing schemes are excellent examples of how architects can create lively communities and affordable housing, using intelligent, sustainable and architecturally innovative designs where our burgeoning urban population are proud and happy to live.

YCONE 住宅楼
YCONE Residential Tower

Ateliers Jean Nouvel

拉康弗伦斯区由让·努维尔设计的新旗舰住宅楼体现该地区的精神面貌
Atelier Jean Nouvel's new flagship residential tower in La Confluence embodies the spirit of the district

YCONE住宅楼位于法国大都市里昂的市中心,是一座由建筑师让·努维尔设计的豪华型公寓楼。YCONE的总占地面积为7150m² (包括92套公寓、650m²的一层服务区以及两层的停车场),拥有双色建筑外表皮、原始的Y形轮廓和令人叹为观止的180°视野,成为拉康弗伦斯区一个新增的大胆设计。

该建筑可为居民提供面积为30m²至215m²不等的公寓,其中65套公寓由VINCI Immobilier销售,27套由Alliade Habitat管理,这一安排为YCONE提供了所有利益相关者所期望的社交组合。YCONE由VINCI Immobilier和Cardinal共同推广,体现了两个实体的互补性;就像拉康弗伦斯区本身一样,这座建筑将大胆的建筑品位与强烈的美学特征有机地结合在了一起。

拉康弗伦斯区位于索恩河与罗纳河的交汇处,这里是里昂的现代区,建筑师让·努维尔的设计理念受到了该地区众多建筑外立面的直接影响。建筑师那独一无二的设计构思与拉康弗伦斯区的千变万化碰撞相遇,诞生了YCONE这一住宅楼建筑。让·努维尔道:"当我设计一个项目时,我经常谈论'拼图中缺失的部分'。在拉康弗伦斯区,YCONE将被三个建筑项目包围,其位置是预先确定的。我试图将建筑物稍微转一圈,推到一侧,然后再将其推到另一侧,以找出与邻近建筑物进行积极互动的方法。但是有关这一点的讨论要基于两个条件:城市化与舒适性。"

YCONE以能够使"两栋建筑合而为一"的楼层移动系统为特征,其主立面是彩色的铝盒,第二层轻质立面由铝构件和带纹理的玻璃构成。这使得立面的外观变得模糊而令人惊讶。努维尔解释说:"多亏了第二层非常轻而且只有一部分的立面,YCONE才得以淡化相似之

处,创造不同之处——在光线、感觉上的不同,当然还有平面上的不同……我在两个平面上设计了立面,并研究了这两个平面之间区域的情况。这两个平面之间的间隙将成为一个居住空间,一个中间区域,日本人称之为'ma'……两个立面叠加形成两个组成部分,进而形成一个稍深的组成部分。"另一个独特的特征是建筑物的Y形轮廓,它是由悬挂的外立面的"鳃"构建的。更加令人瞩目的是顶部的"盖子"——一个重达80t的金属框架,它令建筑更加完整。

主立面的铝盒包含21种不同的柔和色调。努维尔说:"我尝试遵守我的朋友雅克·赫尔佐格和皮埃尔·德·梅隆制定的规范,使用白色符号来标记建筑。(但是)我还是决定,最好使用一些色彩来使老里昂焕发新的生机……我把一点色彩稀释成白色,以便留下一点柔和的痕迹,再通过叠加,来创造几何形状的建筑与色彩的变化。"一个绿色的茧壳结构由梨木和超过10m高的橡木组成,这既可保证居民生活的私密性,又确保了YCONE能成为一个保留自然特色的地方。

YCONE, a luxury apartment building designed by architect Jean Nouvel, is located in the heart of metropolitan Lyon, France. With a total area of 7,150m² (including 92 apartments, a ground floor service area of 650m² and two levels of parking), its dual-colored skin, original Y-shaped silhouette and breathtaking 180-degree view, YCONE is a daring addition to La Confluence district.

Within the residential block, which offers apartments from 30m² to 215m², 65 flats were marketed by VINCI Immobilier and 27 managed by Alliade Habitat, lending YCONE the social mix which was desired by all project stakeholders. Realized in a co-promotion between VINCI Immobilier and Cardinal, YCONE illustrates the complementarity of two entities: just like La Confluence itself, the building respectfully combines a taste for daring architecture with a strong aesthetic signature.

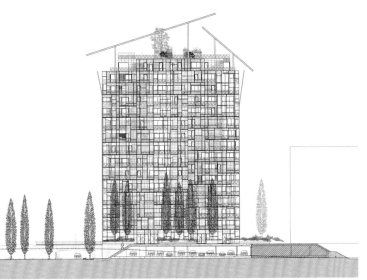

东立面 east elevation

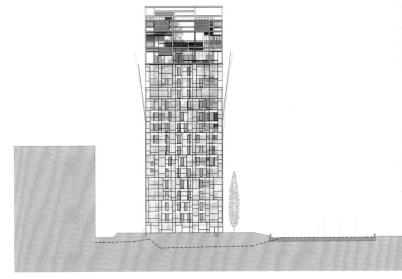

北立面 north elevation

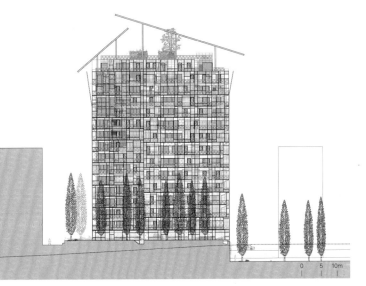

西立面 west elevation

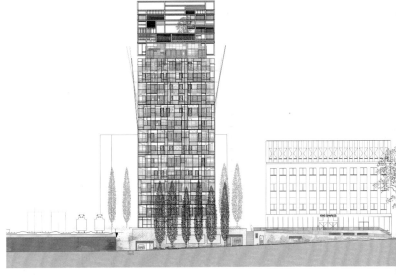

南立面 south elevation

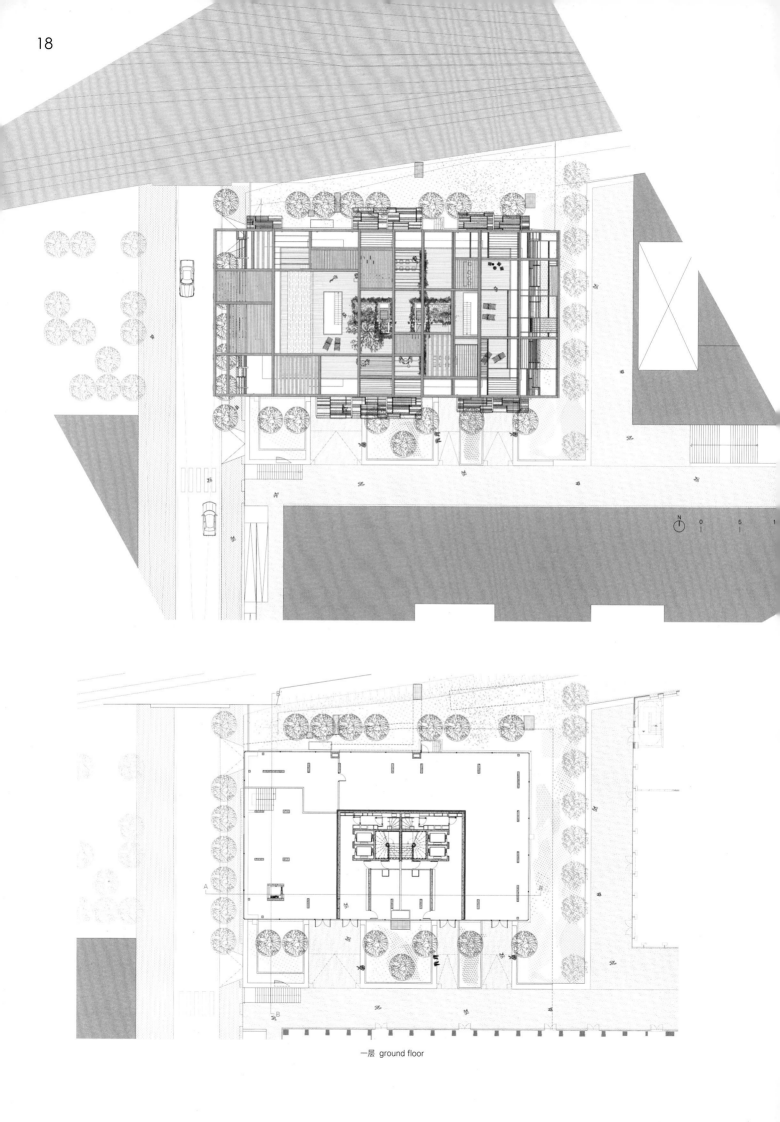

一层 ground floor

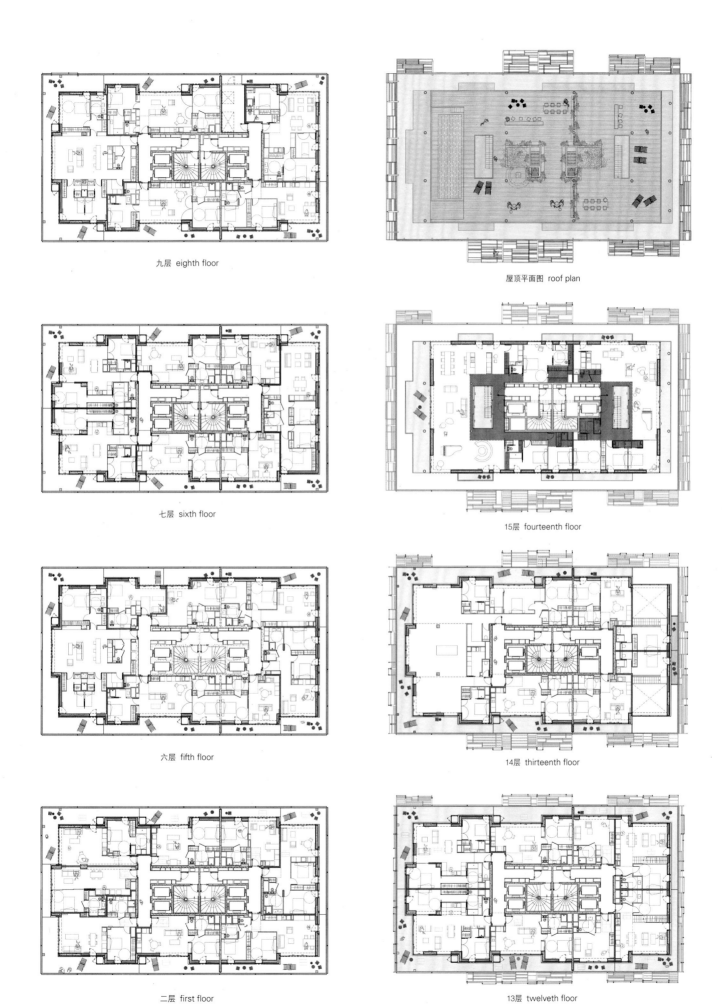

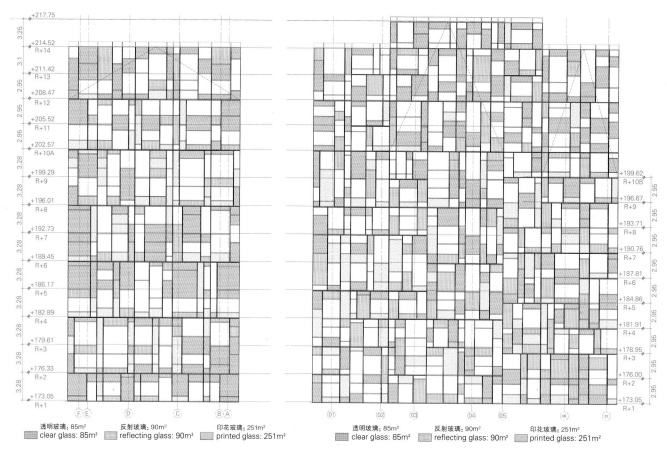

填充玻璃部分的外立面——南立面与东立面
exterior facade-glass filling _ south & east facade

铝盒覆层——南立面与东立面
aluminum cassette cladding _ south & east facade

Architect Jean Nouvel was immediately inspired by the many faces of La Confluence district – the modern district of Lyon where the rivers Rhône and Saône meet. YCONE was born from the meeting between the vision of a creator with the unique and protean place that has become La Confluence. According to Jean Nouvel, "when I'm designing a project, I often talk about 'the missing piece of the puzzle'. In La Confluence, YCONE will be surrounded by three projects and its place is predetermined. I tried to turn the building round a bit, to push it to one side, then push it to the other side, to work out how I could set off a positive conversation with the neighboring buildings. But that discussion was based on two conditions: urbanity and amenity."

With its main facade of colorful aluminum cassettes, and a second lightweight facade of aluminum elements and textured glazing, YCONE is also characterized by a "two buildings in one" floor shifting system. This allows for a blurred and surprising vision of the facade. Nouvel explains: "Thanks to a second facade that's very light and only partial, YCONE plays down similarities and creates differences – in light, feel, and of course, planes… I've designed a facade over two planes and worked on what happens between the two. This gap will be a living space, an in-between area, what the Japanese call ma… The two facade planes will superimpose two compositions that then form one slightly deeper composition." Another unique characteristic is the building's Y shape, created by the "gills" of the overhanging facades. At the top is a "cap", an impressive 80-ton metal frame, which completes the work.

The aluminum cassettes of the main facade comprise 21 different pastel shades. Says Nouvel: "I've tried to respect the specifications laid down by my friends, Jacques Herzog and Pierre de Meuron, playing on the pattern of punctuations in white. [But] I've also decided it would be good to use certain colors that conjure up the old Lyon… I've diluted a drop of color in white so as to leave a trace of pastel that will create variations of color through overlaying, [creating] geometric compositions and chromatic variations." A cocoon of green, composed of pear trees and oaks over 10 meters tall, guarantees the privacy of the inhabitants as well as ensuring that YCONE is also a place reserved for nature.

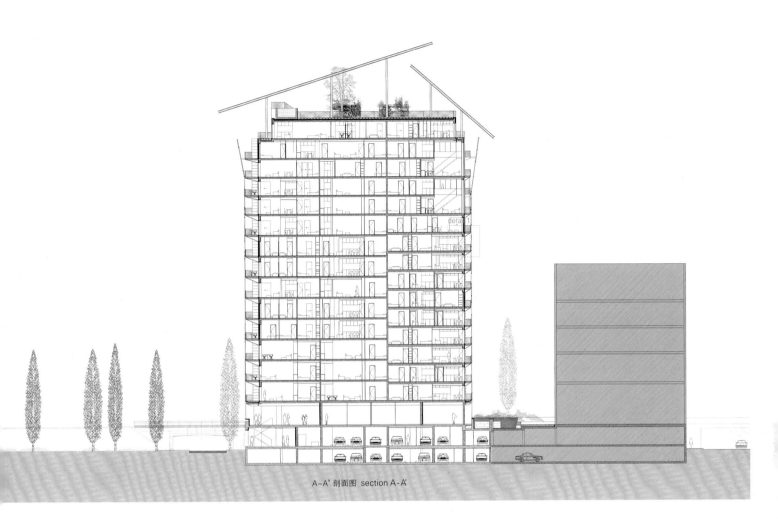

A-A'剖面图 section A-A'

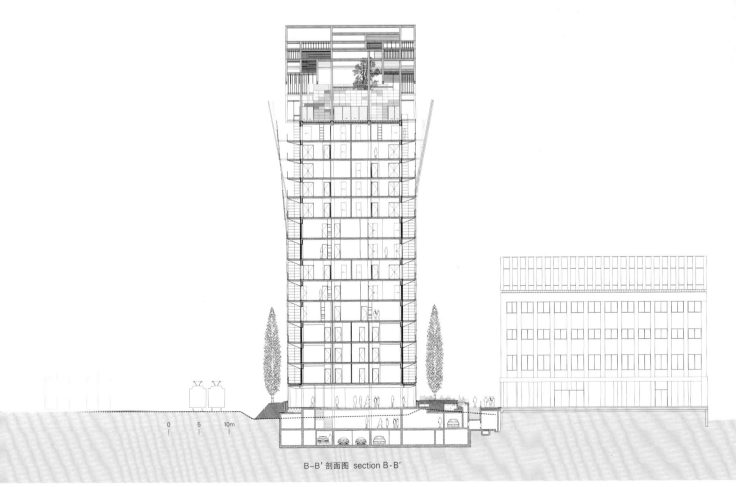

B-B'剖面图 section B-B'

项目名称：YCONE Residential Tower
地点：Passage Panama, Lyon confluence, 2ème arrondissement, France
建筑师：Jean Nouvel – Ateliers Jean Nouvel
顾问：Samuel Nageotte
项目负责人：project – Thomas Amarsy;
project & execution – Alberto Rubin Pedrazzo
项目建筑师：Victoire Guerlay, Rui Pereira,
Marie Charlotte Prosperi, Qiang Zou
实习生：Clarisse Estang, Guillermo Gonzales, Rémi Lapostolle,
Daniel Martinez, Onur Ozman, Giulia Piana
计算机生成图像：Benjamin Alcover, Lionel Arnold,
Mizuho Kishi, Sébastien Rageul, Franklin Tresca
3D模型：Alexandre Braleret, Laura Joo, Jim Rhone
平面设计：Rafaëlle Ishkinazi, Marlene Gaillard,
Eugénie Robert, Vatsana Takham
室内设计：Sabrina Letourneur
景观：project – Laura Giuliani, Jérôme Dureault (assistant);
project & execution – Isabelle Guillauic
工程师：cost consultant – Procobat; structural work – Cogeci;
building services, thermal – Katene; facades – Arcora;
acoustics – Genie Acoustique; sustainable development – Etamine;
execution architectural team – CARDINAL réalisations; security – Socotec
客户：SCCV M3 SUD Confluences
功能：mixed residential tower (social / non-social) – 53-meter high;
ground floor – retails 666m²; 1st~14th floors – residential 5,886m²;
92 apartments from T1 to T5
可用楼面面积：6,552m²
总楼面面积：7,461m²
研究开始时间：2012.10 / 获得建筑许可时间：2013.12.30
施工开始时间：2016.10 / 开放时间：2019.3.7
摄影师：©Roland Halbe

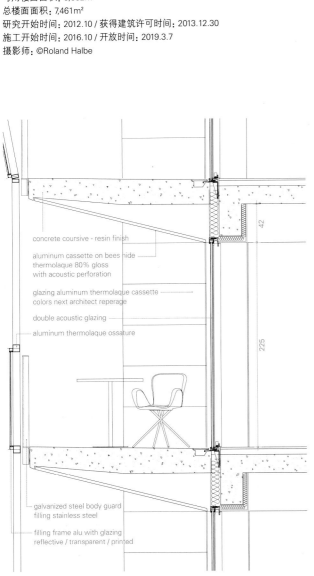

详图1 detail 1

L'Arbre Blanc 公寓
L'Arbre Blanc

Sou Fujimoto Architects + Nicolas Laisné + Dimitri Roussel + OXO Architectes

像树一样的塔楼为城市居民带来温暖与阴凉
The tree-like tower block offers welcome and shade to the city's residents

L'Arbre Blanc公寓（又名白树），是由藤本壮介、Nicolas Laisné、Dimitri Roussel、Manal Rachdi（OXO建筑事务所）以及法国的Marie（藤本壮介工作室的合作伙伴）共同设计建造的。2013年，蒙彼利埃市议会提议发起该建筑项目；"Folie Richter"竞赛旨在确定这座像灯塔一样的公寓的建筑设计蓝图，并期望该项目能在一定程度上丰富城市的建筑遗产。L'Arbre Blanc公寓必须与环境相适应，与城市中的商店和住宅相协调。

尽管建筑师们获得设计灵感的源泉（即大自然）是相同的，但是他们通过不同的方式展现了自己的设计理念，因此该建筑在设计上展现出了丰富的形式。

建筑师专注于人性化维度，在建筑的底部和顶部创造了公共空间。一层是一个通向街道的由玻璃墙围合的空间，屋顶则有一个向公众开放的酒吧和一个供居民使用的公共区域。这样，即使是住在公寓二层的住户也可以欣赏高处的风光。

该项目与众不同的点在于它独特的树状设计。阳台从树干上延伸出来，投下的阴影保护着建筑的外立面。阳台和凉廊的设计改善了居民的户外活动，并增加了居民之间的联系方式。每间公寓均拥有至少7m²（最大面积可达35m²）的室外空间，具有多层次的私密和布局选择。为了让所有住户都能欣赏到令人愉悦的景色，建筑师们利用一系列空间实验以及物理三维模型来制作设计蓝图。

L'Arbre Blanc公寓的设计运用了许多创新性技术。例如，露台的设计。露台的悬臂长达7.5m，居世界第一。这些独特的户外空间成为功能齐全的客厅，与室内空间相连，使居民能够灵活运用室内和室外两个空间，这对一年中只有80%的时间沐浴在阳光下的城市来说是一件多么奢侈的事情！

阳台的设计就像彼此穿插着寻找阳光的树叶一样，这一设计比例强调了充分利用户外空间的目标。宽敞的阳台也是对解决"南方生态"这一环境问题的回应。延伸出来的阳台成为外立面有效的防护面纱，提供必要的遮阳效果，还可以分散侧面刮过来的风，让空气更加顺畅地流通。

建筑师为这个综合用途开发项目中的塔楼生活设计了一种新的方法。为了避免造成"令人无法接近的塔楼综合征"，从最初的讨论开始，建筑师就将重点放在开发公共空间上，决定沿勒兹河扩建一座景观公园，并使公寓楼向公众开放。

这座17层高的建筑完全融入了城市生活之中，向蒙彼利埃市的所有人开放。建筑一层设计有一个艺术画廊，屋顶酒吧与全景花园相连。人们可以真正拥有这座塔楼，它将成为蒙彼利埃人民的骄傲，同时也成为这座城市的旅游景点。

L'Arbre Blanc (the White Tree) is the result of a collaboration between Sou Fujimoto, Nicolas Laisné, Dimitri Roussel and Manal Rachdi of OXO architects, with Marie de France, partner of Fujimoto's Paris studio. The project was initiated by Montpellier city council in 2013; the "Folie Richter" competition sought to identify a blueprint for a beacon tower to enrich the city's architectural heritage, which must fit into its environment and must also incorporate shops and homes.

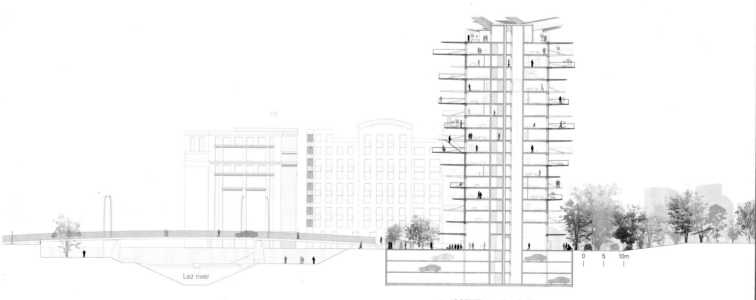

A-A' 剖面图 section A-A'

Each of these architects seek inspiration in nature even if they express it in different ways; this led to mutually enriching outcomes.

The architects focused on the human dimension, creating public spaces at the bottom and top of the building. The glass-walled ground floor opens out onto the street, while on the roof there is a public bar and a common area for residents, so that even the owners of first-floor apartments can enjoy the view.

What sets the project apart is its tree-like design, with balconies that branch off the trunk and shades that protect its facade. These balconies and pergolas promote outdoor living and enable a new type of relationship between residents. Each apartment boasts an outdoor space of at least 7m^2 (the largest is 35m^2) with multiple levels of privacy and layout options. So that all apartments have pleasing views, the architects sculpted the blueprint with a series of spatial experiments using physical 3D models.

The many technical innovations of L'Arbre Blanc include the terraces, whose cantilevers of up to 7.5m long constitute a world first. These exceptional outside spaces are fully-fledged living rooms which are connected to the dwellings in such a way as to allow residents to live both inside and outside – a luxury for a city bathed in sunshine 80% of the year!

The proportions of the balconies emphasize this aim to embrace the outdoors, as do the leaves that fold out in search of the sunlight. These generous balconies are also a response to the need for environmental solutions closely tailored to the "ecology of the south". Forming an effective protective veil for the facade, they provide the necessary shade and break up skew winds to help air circulate more harmoniously.

The architects adopted a new approach to tower living for this mixed-use development. To avoid "inaccessible tower syndrome", from the earliest discussions, there was a real focus on public space, including extending a landscaped park along the Lez River and opening the tower up to the public.

The seventeen-story building is a full participant in city life, aiming to be accessible to the people of Montpellier, with an art gallery on the ground floor and a rooftop bar linked to a panoramic garden. By allowing people to take physical ownership of the tower, it will become an object of pride for the people of Montpellier, as well as a tourist attraction.

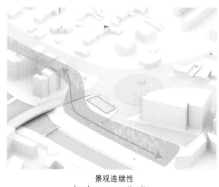

景观连续性
landscape continuity

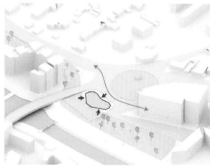

整合与城市动线
integration and urban movement

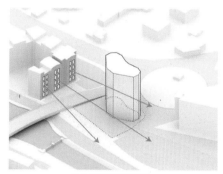

适应并尊重环境的建筑
a building adapted and respectful of its environment

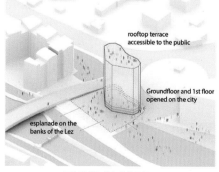

面向勒兹河并拥有全景视野
opening on the River Lez and on the panorama

捕捉光线，降低风速
catching the light and slowing down the wind

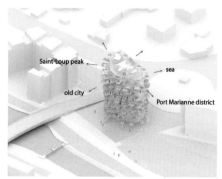

户外生活并可以欣赏无可比拟的全景景观
living outside and enjoying unrivaled panoramas

17层 sixteenth floor

屋顶平面图 roof plan

十层 ninth floor

一层 ground floor

体块研究
mass study

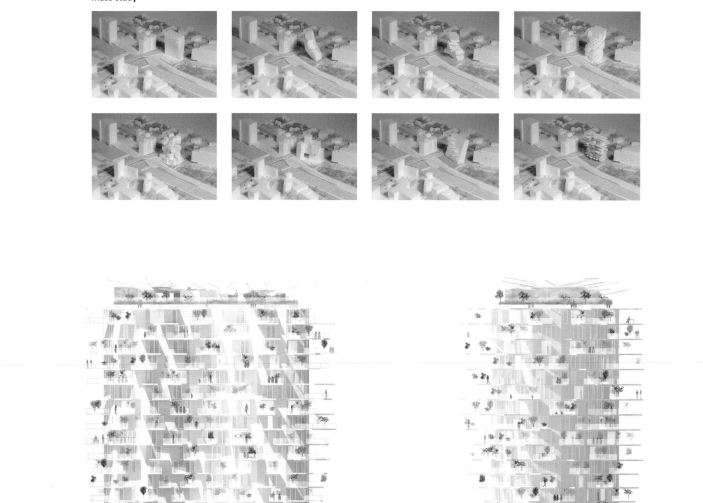

东立面 east elevation

南立面 south elevation

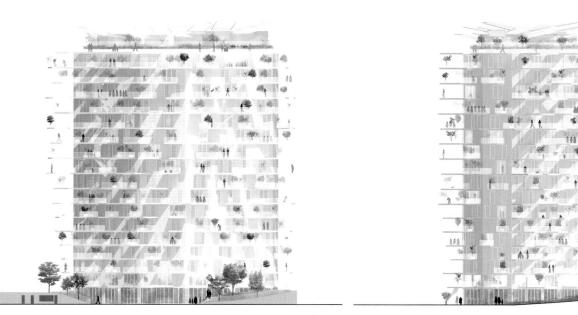

西立面 west elevation

北立面 north elevation

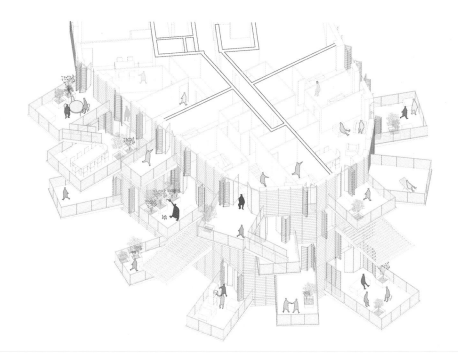

项目名称：L'Arbre Blanc
地点：Place Christophe Collomb, 34000 Montpellier, France
事务所：Sou Fujimoto Architects, Nicolas Laisné, Dimitri Roussel, OXO Architectes
项目管理：Marie-Laure Coste-Grange
施工管理：CAP Conseil, icK
景观设计师：Now Here Studio
工程师：structure - André Verdier; fluids – ARGETEC; environmental – Franck Boutté Consultants; cost management – VPEAS; surveying, roads and services – Relief GE; lighting - Les Eclaireurs; inspection – SOCOTEC; fire performance – Efectis
分包商：Carcass – Fondeville; steel structure – Languedoc Etanchéité, SPCM; facades – CIPRES; electricity – ENGIE; HVAC – Midi-Thermique
客户：Opalia, Promeo Patrimoine, Evolis Promotion et Crédit Agricole Immobilier Languedoc-Roussillon GSA Réalisation (Delegated contractor)
建筑面积：10,225m²
造价：€20.5m before tax
竞赛时间：2019
摄影师：
Courtesy of the architects - p.32~33, p.37, p.38
©Cyrille Weiner (courtesy of the architect) - p.31, p.39, p.41

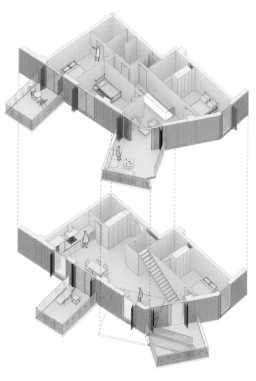

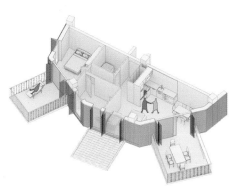

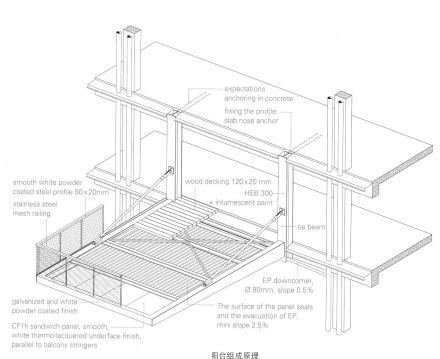

阳台组成原理
principle of composition of balconies

La Borda 合作住房
La Borda Cooperative Housing

Lacol

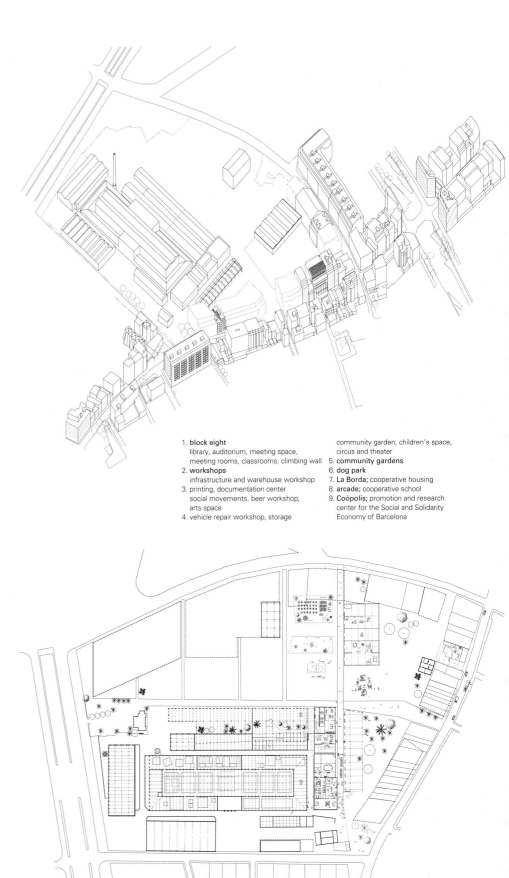

1. **block eight**
   library, auditorium, meeting space, meeting rooms, classrooms, climbing wall
2. **workshops**
   infrastructure and warehouse workshop
3. printing, documentation center
   social movements, beer workshop, arts space
4. vehicle repair workshop, storage
   community garden, children's space, circus and theater
5. **community gardens**
6. dog park
7. **La Borda**; cooperative housing
8. **arcade**; cooperative school
9. **Coòpolis**; promotion and research center for the Social and Solidarity Economy of Barcelona

Lacol 事务所设计了一个社区友好型且可持续的社会保障性住房
Lacol designs a community-friendly and sustainable social housing

La Borda合作住房是由Can Batlló社区推动的一个建筑开发项目。该项目是社区居民自发提出的,旨在让人们获得体面、非投机的且经济实惠的住房。

打造合作住房的想法始于2012年,同时还要修复巴塞罗那桑茨地区附近的工业建筑。

该建筑项目位于公共用地内,它所包含的社会保障性住房的租期为75年。它处在Constitució街道,位置与Can Batlló工业区相连,但其中的一个立面面向La Bordeta的原有社区。

项目开始之初,Lacol建筑师合作事务所就参与其中,为巴塞罗那严重的住房危机提供了可供选择的解决方案。这样的参与方式成为一种机会,让人们可以自下而上重新思考人们欢迎的住房的建造方式,同时还可以让未来的用户都参与进来。

重新定义集体住房项目

La Borda合作住房的理念有利于促进社区关系的良好发展,进而可以使居民通过社区空间增进彼此之间的关系,在家庭工作和护理领域建立合作关系。这是通过使某些私密的日常生活变得可见而实现的。

该项目计划建设28个住房单元(面积分别是40m²、60m²和75m²)和社区空间,这样,人们日常生活的方方面面都可以从私人空间延伸到公共空间,增强彼此联系,进而改善社区生活。此类社区空间包括厨房-餐厅、洗衣设施、客房、健康和护理空间、存储空间以及诸如天井和屋顶等室外空间。这些空间围绕着一个大型中央庭院连接起来,进一步加强了居民之间的相互联系。这样的建筑格局让人想起曾经在西班牙中部和南部流行的住宅类型"corralas"。

可持续性与环境质量

建筑师的目标是以最低的能耗实现住房最大的舒适性,并且对环境的影响达到最小,如此便可以通过降低能耗来减少用户们的住房成本。设计师坚信,最好的策略是减少建筑对所有环境因素(能源、水、材料和废物)的初始需求,因此应该优先考虑被动式能源策略,以最大限度地利用现有资源。

为此,该建筑共有六层结构使用了交叉层压木。这种轻质、优质、可再生的材料具有较低的能源成本。目前,La Borda公寓是西班牙最高的交叉层压木结构建筑。

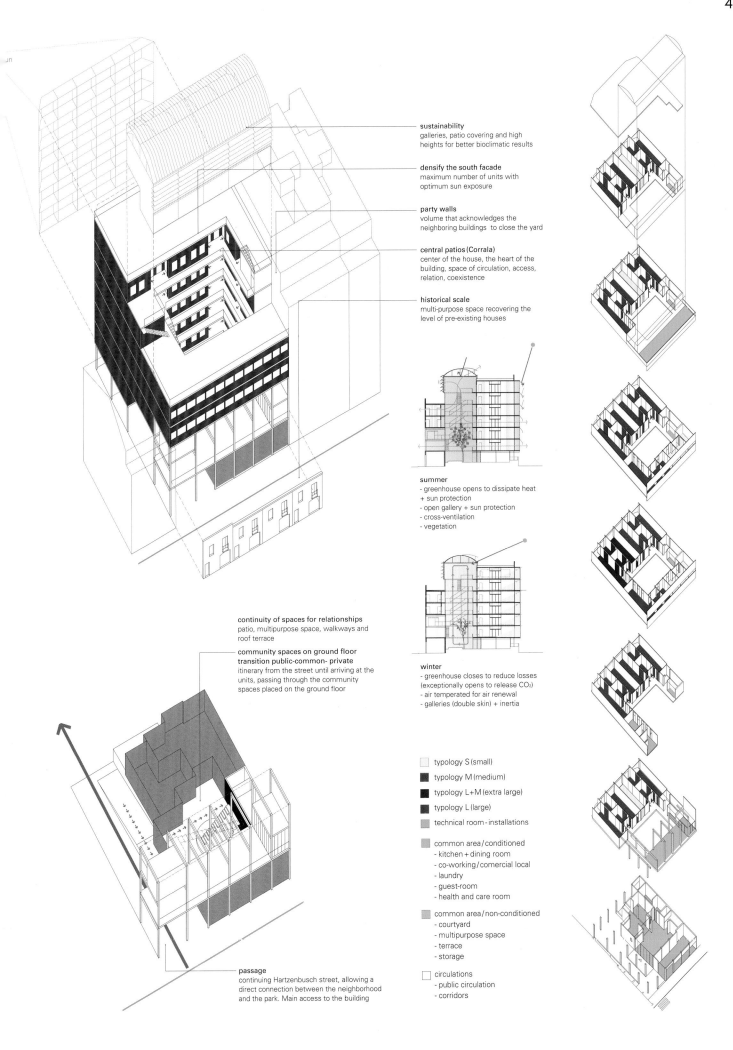

用户参与

未来用户对整个(设计、建造、使用)过程的参与是本项目最重要、最独特的变量。

在设计过程中,用户的参与是通过建筑委员会进行的。该委员会是技术团队和用户之间的纽带,也是负责筹备建筑工作室的一方。

住房负担能力

合作住房的一个必不可少的条件是保证低收入居民获得体面且经济实惠的住房。建造成本是衡量每月租金价值的一个决定性因素。为了尽量减少初期投资,将该建筑项目分成两个阶段。第一阶段达到了允许居民使用该建筑的最低居住要求;第二阶段将允许社区随着时间的发展完成剩余的项目开发。

La Borda cooperative housing is a development driven by the community of Can Batlló – a self-organized initiative to access decent, non-speculative, affordable housing.
The idea for the cooperative housing began in 2012, with the recovery of industrial premises in the Sants neighborhood of Barcelona.
The project is located on a public land, and comprises social housing with a 75-year lease. The properties are located on Constitució Street, in a position which borders the industrial area of Can Batlló, but with a facade that looks toward the existing neighborhood of La Bordeta.
Lacol, the cooperative of architects involved since the beginning, was motivated to contribute an alternative solution to the serious housing crisis in Barcelona. This engagement became an opportunity to rethink the production of popular housing from the bottom-up and with the participation of future users.

Redefining the Collective Housing Program

The concept of La Borda is to promote community-friendly forms of existence, enhancing the relationships between resi-

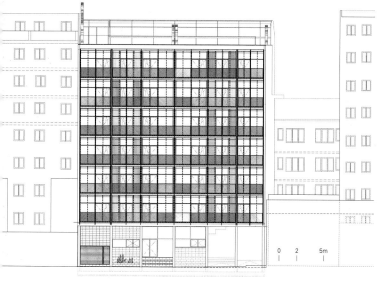

南立面 south elevation

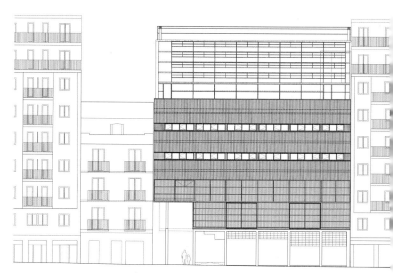

北立面 north elevation

dents through community spaces, and establishing cooperation between them in the fields of domestic work and care. This is achieved by making certain private spheres of everyday life more visible.

The building program proposes 28 units (of 40, 60 and 75m²) and community spaces that allow for aspects of everyday life to be stretched from the private into the public spheres, thereby enhancing community life. Such spaces include the kitchen-dining room, laundry facilities, guest areas, spaces for health and care, storage, and exterior spaces such as patios and roofs. All of these are articulated around a large central courtyard, which further encourages relationships between residents. This is reminiscent of the corrala – a popular typology in central and southern Spain.

Sustainability and Environmental Quality

The architects' objective was to build with the lowest environmental impact and to achieve comfortable homes with minimum consumption. This would reduce the costs of access to housing by eliminating energy poverty among users. The designers had a conviction that the best strategy would be to reduce the initial demand of all the environmental vectors (energy, water, materials and waste), prioritizing passive strategies to achieve maximum use of existing resources.

The structure of six floors is of Cross Laminated Timber (CLT), a lightweight, high quality, renewable material that has a lower energy cost. La Borda is currently the tallest CLT-structured building in Spain.

User Participation

The participation of the future users in the whole process (design, construction and use) is the most important and unique variable of this project.

During the design, participation was articulated through the architecture commission, which was the link between the technical team and the general assembly, and the party responsible for preparing architectural workshops.

Affordability

An indispensable condition of the cooperative was to guarantee access to decent and affordable housing as an alternative for low income residents. The cost of construction was a determining factor in establishing the value of the monthly rental fee. The project went through two phases to minimize this initial investment. The first achieved the livable minimums that allow residents to use the building; the second phase will allow the community to complete the remaining developments over time.

五层 fourth floor　　　　　　　　　　　　　六层 fifth floor

项目名称：La Borda Cooperative Housing / 地点：Constitució 85-89, Barcelona, Spain / 事务所：Local Arquitectura Cooperativa / 参与者：Arkenova, Miguel Nevado, AumedesDAP, Societat Orgànica, PAuS (Coque Claret, Dani Calatayud), Grisel·la Iglesias (Àurea acústica) / 客户：La Borda / 建筑面积：3,000m² / 设计时间：2014 施工时间：2017.2—2018.10 / 摄影师：courtesy of the architect - p.42~43, p.44, p.52~53; ©Lluc Miralles (courtesy of the architect) - p.47, p.50, p.51, p.55[lower]; ©Gabriel Lopez (courtesy of the architect) - p.55[upper]

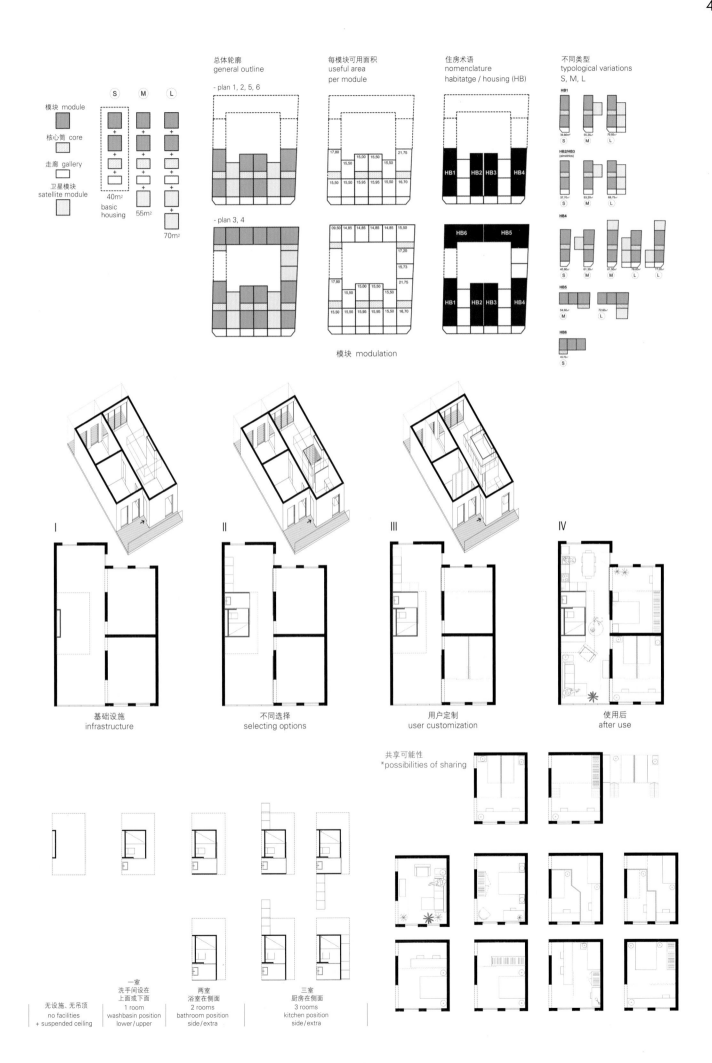

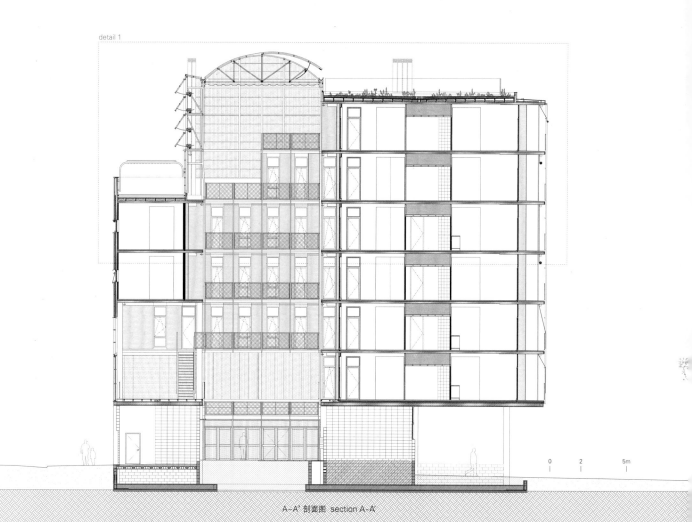

A-A' 剖面图 section A-A'

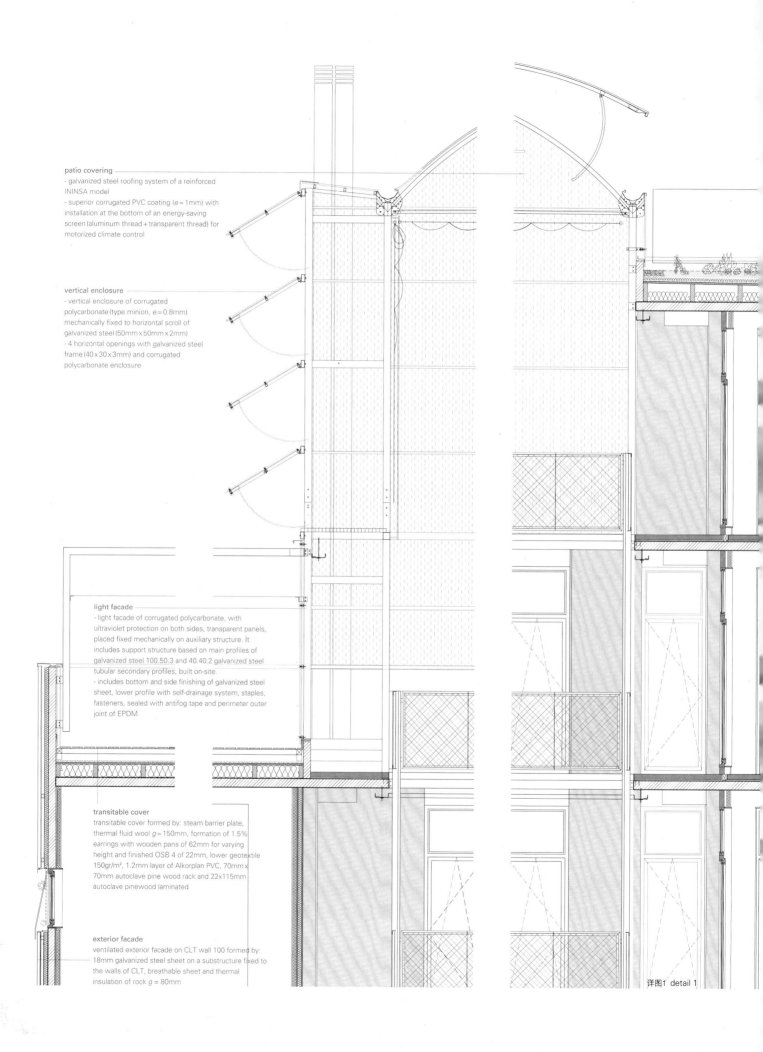

vegetation cover
vegetation cover formed by: steam barrier plate, thermal insulation of rock wool g = 150mm, 1.5% slope formation with 62mm wooden pine slats for variable height and OSB 4 of 22mm, lower geotextile 150gr/m², 1.2mm layer of Alkorplan PVC, protective blanket and retainer SSM45, drainage plate FD 25, filtering sheet SF, substrate ZinCo Terra Sedum 10cm.

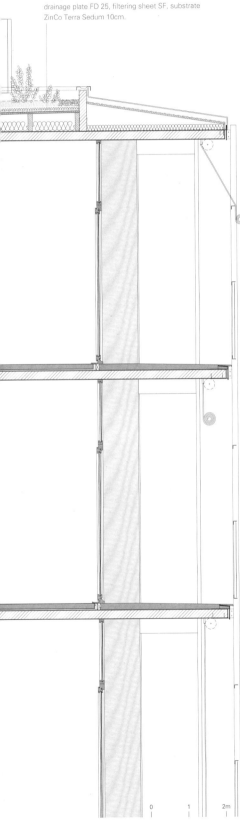

公共建筑与空间营造

Public Bu
and Plac

长久以来，我们因公共建筑的作用、范围和功能及其对某一地方、某个国家或特定社会的影响而对其进行定义。它们曾经代表着神圣、崇高、特权和荣耀，而如今的公共建筑似乎更倾向服务于人文价值，通过提供一种"空间感"，让人们在同一屋檐下实现空间共享和交流互动。这种空间的建造过程绝非一成不变，也不仅仅追求审美效果；相反，它在很大程度上取决于区域政策、人文地理、社会一致性、城市拓扑结构以及意识形态、文化、遗产、象征性表现等因素。因此，建筑只是

Public buildings have consistently been defined by their roles, scopes, and functions, as well as the impacts they have on a place, a state, a country, or a society. Once a representative manifestation of the divine, the sublime, or the prerogative and the glory of an authority, today's public buildings seem to aim more toward serving humanistic values by providing a "sense of place" that allows people to share space and interact under their shed. This placemaking process can hardly be a single aestheticized act of architecture; instead it is largely dependent on issues of regional policy, human geography, social coherence, urban topology, as well as ideology, culture, heritage, and

绿色广场图书馆和广场_Green Square Library and Plaza / Studio Hollenstein
de Maio街24号SESC综合体_SESC 24 de Maio / Paulo Mendes da Rocha + MMBB Arquitetos
MÉCA文化中心_MÉCA Cultural Center / BIG
比奥比奥地区剧院_Biobío Regional Theater / Smiljan Radic + Eduardo Castillo + Gabriela Medrano
高雄艺术中心_Kaohsiung Center for the Arts / Mecanoo architecten
公共建筑与空间营造_Public Buildings and Placemaking / Angelos Psilopoulos

在全局方案中发挥自身作用,而非凭借一己之力控制公共建筑的范围和视野。
　　然而,本文中所提到的建筑案例表明,建筑可以通过发挥自身优势来应对这些复杂的问题:阅读字里行间的意思,并以最具洞察力和创造性的方式解释一个特定的设计。通过这种方式,要想使建筑能吸引社区居民的眼球,首先要使其成为一个能吸引眼球的地方。

symbolic representation. Therefore, architecture merely play a part in a global scheme of things rather than hold upon itself alone the scope and vision of the public agents that commission these buildings.
However so, the collection of buildings in this article show us that architecture can reply to these complex problems by doing what it usually does best: reading between the lines and interpreting a given program in an insightful and most creative manner. In this way, the need to make the building an attractor to the community is answered simply by making it an attractive place to be.

空间营造的表达形式——公共建筑
Public buildings as expressions of placemaking

Angelos Psilopoulos

　　建筑和公共领域之间的关系十分微妙,它小心地游走于建筑美学和其在更广泛的社会尺度上所产生的影响之间。然而,在建筑历史上的大部分时间,问题都比较简单,因为人们赋予了建筑物以独特的视角,比如神圣的美德、皇家的特权、私人的高贵、社会的改革,或者强调崇高性。这一特点为新古典主义和现代主义所共有。不过,在20世纪70年代早期的"grand maîtres"式委托任务衰落之后,一种新的公共建筑类型取代了它的位置,[1]而且这种新类型是最不关注象征主义的建筑类型之一。事实上,伦佐·皮亚诺、理查德·罗杰斯和苏·罗杰斯以及詹弗兰科·弗朗西尼设计的法国巴黎蓬皮杜中心成为一系列公共建筑的先驱,这些建筑从未像现在这样形式多样、用途广泛、有趣生动。因其与众不同的特点,这些建筑更倾向于与其所处的城市框架相得益彰,而非突出其在周围环境中的自我参照性。20世纪末的大部分时间里,法国"弗朗索瓦·密特朗的宏伟计划",弗兰克·盖里的毕尔巴鄂古根海姆博物馆,或者巴塞罗那市的1992年奥运会维拉·奥林皮卡·德尔·波布卢努项目以及毗连建筑项目的大胆尝试,都坚定了市政当局对市政建筑力量的信念,这种力量能够在国际范围内大规模重塑公共领域,重建场所和空间。

　　有观点认为当代公共建筑是对公共领域的一种改造,而非一种单一的建筑行为。20世纪80年代初,阿尔多·罗西写的《城市建筑》一书具有深远影响,该书清楚地介绍了以上观点。根据罗西的观点,从城市设计的角度来分析建筑,我们可以将其看作是城市结构的重塑,从而创造出文化传承和集体记忆留存的新思路。因此,罗西提出建筑应该关注建造过程,包括诸如"城市地理、城市地形、建筑风格和其他相关学科"等方面。[2]几十年后,伊格纳西·德·索拉·莫拉雷斯提出,"随着对科技进步的信心丧失,以及该学科价值观支离破碎",应该将这种类型的建筑作为一种尝试来分析,"这种尝试努力建设的并非具体的建筑,而是在价值观念莫衷一是、变化不定,以及充分有力的基础参考匮乏时,提供一种借鉴意义",[3]即试图从一个比维特鲁威的功利主义、坚定主义和美学价值等更深远的维度中推断建筑的功能和特性。[4]实际上,"宏伟计划"的经验毫不含糊地向我们表明,以上大胆尝试

The relationship between architecture and the public realm has always had to walk a very thin line between a building's aesthetics and the impact it bears on a wider societal scale. Nevertheless, for the most part of architectural history, things were simple as buildings were invested with singular visions such as divine virtue, royal prerogative, private nobleness, social reform, or an iteration of the sublime. This characteristic was shared by neoclassicism and modernism alike. However, since the decline of the "grand maîtres" type of commission in the early 70's, a new genre of public architecture has taken its place;[1] and it is one that the least of its preoccupations lies in symbolism. Indeed, the realization of the Center Georges Pompidou by Renzo Piano, Richard and Sue Rogers, and Gianfranco Francini, in Paris, France, spearheaded a series of public buildings that have since never been more diverse, versatile, playful. For their most distinctive trait, these buildings prefer to develop in thoughtful dialogue with the civic framework they occur upon, instead of imposing a self-referring character on their surroundings. For the most part of the late 20th century, bold ventures such as François Mitterrand's Grands Projets in France, Frank Gehry's Guggenheim Bilbao, or the City of Barcelona's development of the Vila Olímpica del Poblenou and adjacent projects for the 1992 Olympic Games solidified civic authorities' faith on the power of civic architecture to disrupt the public realm and regenerate place and space on a scale of international scope and ambition.

The claim that contemporary public buildings lay on restructuring the public realm is hardly a singular act of architecture. This idea is clearly introduced in Aldo Rossi's seminal book *The Architecture of the City*, written in the early '80s. According to Rossi, looking at buildings from an urban design standpoint allows us to examine them as disturbances in the fabric of the city that create new strands of cultural continuity and collective memory. Thus Rossi proposes that architecture should be a matter of process, comprising of aspects such as "(…) urban geography, urban topography, architecture, and several other disciplines".[2] A couple of decades later, Ignasi de Solà-Morales proposes that "with the loss of confidence in scientific and technological progress and the values of the subject shattered", this type of architecture should be analyzed "as endeavors to construct not a building but a meaning that reflects the precariousness of (…) the dispersion of values and the absence of unshakable, foundational referents",[3] namely as attempts to extrapolate function and identity out of a context that spans remarkably further than Vitruvius' Utilitas, Firmitas, and Venustas.[4] Indeed, the experience of the "Grands Projets" et al. showed us in no uncertain terms that the outcome of these ventures was also largely due to matters of policy, regional geographies, micro and

的结果很大程度上取决于政策问题、区域地理位置、微观和宏观经济学甚至广告等因素,它还提出了社会一致性、意识形态、文化、遗产和象征性解释等诸多问题。

然而,机构建筑作为逃离日常生活琐事的避难所,以及特定时期高雅文化的堡垒,它们始终如一地发挥着作用。同样,机构建筑作为集会场所和社会改革的载体,能使公众接触崇高的理想,如公民美德、社会礼仪或对艺术的欣赏。尽管如此,当今它们的代表作用逐渐消失,转而变成一个更容易为大众接近的形式,这一形式主要解决文化表达的包容性和多样性以及建立日益多样化社会的必要性问题。[5] 尽管权力和政治色彩仍然存在,但这种新建筑风格对其提出的空间和形式操控论义正词严,并且敢对既定的指示提出挑战,在这方面,我们不得不称赞它的勇气,因为它不是对权威资本化愿景的赞颂,而是对在社区中集会和相聚这一由来已久的权利进行创造性重新诠释——就像古希腊"广场"的概念一样。

有趣的是,我们展示的所有项目都利用建筑的公共性来达到修复目的:一方面,旨在支持日常生活并在当地创建社区;另一方面,通过它们的重塑力量来修复或重新利用一块具有潜藏价值的土地,无论贫瘠还是肥沃,无论是融入城市的整体结构还是向着新的方向扩展。考虑到其目的是为公众服务,一些催化式的干预行动大多是良性有益的。当然,将单一的建筑命题作为"城市催化剂"的想法并非没有错误。[6] 对于每一个"蓬皮杜中心",我们都可能会在社区层面破坏当地功能和活动,这主要是由新情况吸引下外来人口的突然涌入所引起的。对于每一个"毕尔巴鄂古根海姆",我们必须考虑其独特的存在会影响城市结构的连续性。随着时间的推移,这些都是我们的建筑将要面临的合理的迫切要求。目前,我们的重点仍然放在设计意图上,正如一位作家在一部充满无望乐观主义的电影中所说的那样:"如果你建立了它,它们就会到来。"[7]

另一方面,无论是在本地范围还是在全球领域,成为亮点似乎是所有建筑吸引眼球都要有的重要策略。为了增强吸引力,建

macroeconomics, and even advertisement, while it also raised issues of social coherence, ideology, culture, heritage, and symbolic interpretation.

However so, institutional architecture acted consistently through the ages as a refuge from the trivialities of everyday life, as well as a bastion for the high culture at a given time and era. Similarly, its buildings acted as places of assembly and served as a vehicle for social reform by exposing the public to high ideals such as civic virtue, social etiquette, or the appreciation of the arts. Still, their representational role gradually dissolved in recent times into a more accessible version, largely addressing issues of inclusion and variety of cultural expression as well as the need to appeal to an increasingly diverse society.[5] This new crop of architecture rather comes off as unapologetic for the spatial and formal manipulations it proposes, as well as defiant for the interpretations it gives to the given brief, albeit the implication of power and political expression still remains. In this respect, we cannot but praise its boldness, for it stands not as a tribute to the capitalizing vision of an authority but rather as a creative reinterpretation of the age-old prerogative to assemble and to be together in civic community – much like as in the ancient notion of "Agora".

Interestingly, all of the projects we present use their publicness towards a healing purpose: on one hand by aiming to support everyday life and create community at a local level; and on the other, by using their disruptive power to revive or repurpose a latent piece of land, either hollow or dense, integrated in the city's fabric or expanding the city to a new direction. Given that they aim to serve the public interest by definition, these ventures of catalytic intervention come off mostly like a benign practice. Of course, the idea of referring to singular architectural propositions as an "urban catalyst" has not been without criticism.[6] For every "Center Pompidou", we stand in danger of creating a disruption of local functions and activities at a neighborhood level, mostly raised by the sudden influx of outsiders that are drawn to the new situation. For every "Guggenheim Bilbao", we have to consider the isolating effect of its singular presence against the continuity of the city's fabric. These are all valid preoccupations that our buildings will be put against in the passing of time. For the moment, the emphasis remains on intentions: "If you build it, they will come", as one writer puts it in a movie imbued with hopeless optimism.[7]

On a different note, being an attraction seems to be the one power all these buildings share primarily as they seek to bring people in, either at a local scale and/or at a global scale. And in order to nurture an invitation, they seem to rehash the time-tested strategies of the recent past, albeit reflecting contemporary concerns.

筑似乎重新使用了过去几年久经考验的战略，然而这些战略反映出了当代的担忧。

　　Bjarke Ingels Group (BIG) 设计的波尔多MÉCA文化中心（92页）模仿巴黎拉德芳斯拱门的风格。然而，它不是一个充满象征意义的框架，反而更像是一条活跃的行人通道，在倾斜的阶梯、坡道和体量的几何结构中促进人们的活动。斯图尔特·霍伦斯坦恩建筑师事务所仿照下沉广场和法国著名的国立图书馆设计的悉尼绿色广场图书馆和新区绿色广场的购物中心（62页），其适宜步行的城市广场的规模创造了更多的上空间，为环绕在它周围的密集的社区带来了舒缓，令人愉悦。Paulo Mendes de la Rocha & MMBB建筑事务所设计的圣保罗de Maio街24号SESC综合体社区中心（74页）将蓬皮杜中心遗产作为一种战略性介入措施，很好地融入了原有的城市结构，并解决了遗产建筑的问题。不过，建筑师并没有拆除旧址，而是提出了一个创造性的缜密再利用方案，既适应场地又适应原有建筑的结构，还可以为城市的需求营造空间。由Mecanoo建筑事务所设计的台湾地区高雄艺术中心（128页）反映出SANAA的劳力士学习中心的设计，是开放公园中的一个充满活力的元素。然而，它并不是一个由中空间组成的景观，而是在密集的高楼之间形成了高低起伏的格局：框架的运用和开放的棚屋，有组织的活动和自发的聚会，所有这些都通过简单地模仿当地土生土长的榕树的宜人树荫而融合在一起。最后一个例子是智利康塞普西翁由Smiljan Radic, Eduardo Castillo和Gabriela Medrano设计的比奥比奥地区剧院（110页），与毕尔巴鄂古根海姆博物馆类似，在一片毫无特色的区域一枝独秀，吸引着人们的注意。虽然它的规模不大且并非富丽堂皇，但其夸张的特点替代了既定的规矩，提供了一种顽皮的"幕后"体验。

　　因此，这些建筑都是基于同一种诠释，它们的设计可能确实是问题驱动的，而不仅仅"痴迷于其物理形式的印象和美学"。[8] 当然，建筑学总是热衷于从赋形中提取价值，而我们本文提到的几个案例也几乎是这样做的。另一方面，由于它们的功能主要目标之一是吸引人流，因此它们的设计是完全合理的，建筑首先应考虑的就是以一种有吸引力的方式呈现。在这个框架中，问题驱动的方法并不包含对美学的否定。但这些建筑不能仅仅是美学意义上的建筑；相反，它们的美在于能够实现人与空间、人与地方、人与

The MÉCA Cultural Center in Bordeaux by Bjarke Ingels Group (BIG) (p.92) follows in the footsteps of the Parisian Arche de la Défense; yet it is more of an active pedestrian passage than a symbolically charged frame, fostering activities in its skewed geometry of steps, ramps and volumes. The Green Square Library and Plaza (p.62) in the new district of Green Square, Sydney, by Stewart Hollenstein Architects echoes the sunken yards and prominent volumes of the Bibliothèque Nationale de France, yet its intimate scale creates more of a walkable city plaza whose void space offers a most welcome relief to the dense neighborhood projected to surround it. The SESC 24 de Maio community center (p.74) in São Paulo, by Paulo Mendes de la Rocha & MMBB Arquitectos takes after the legacy of the Center Pompidou as a strategic intervention embedded well within the existing city fabric, as well as addressing the issue of heritage building; however, instead of a demolition the architects propose a creatively thoughtful reuse that adapts to both the site and the existing building's structure, arranging space by following the city's needs. Kaohsiung Center for the Arts in Taiwan, by Mecanoo architecten (p.128) mirrors SANAA's Rolex Learning Center as a revitalizing oddity among an open park; yet, instead of a landscape of voids it undulates between densities: shelled uses and open sheds, defined activities and spontaneous gatherings, all bound together by the simple evocation of the accommodating shade of the indigenous banyan tree. Finally, the Biobío Regional Theater in Concepción, Chile, by Smiljan Radic, Eduardo Castillo, and Gabriela Medrano (p.110) resembles the Guggenheim Bilbao in acting as a singular point of interest to an otherwise blunt area. However, its modest scale shares no aspirations of grandeur while its theatrical character playfully exchanges the established etiquette for a mischievous "behind-the-scenes" experience. Therefore, these buildings are all based on an interpretation, and their design may indeed be problem-driven instead of merely being "obsessed with the impressions and aesthetics of their physical form".[8] Of course, architecture has always been keen to extract value from form-giving, and our specimens hardly refrain from doing that as well. On the other hand, since one of the principal requirements of their program is to act as attractors, it makes perfect sense that they should look first and foremost for a way to be attractive. In this framework, a problem-driven approach doesn't entail the negation of aesthetics. But these buildings cannot be architecture in mere aesthetic terms; instead, their beauty lies in the ability to constitute a place of liaisons, between the people and the space, the people and the place and the people amongst themselves. In the end, the publicness of these buildings lies on whether they manage to foster assembly. And in order to stay relevant, their architects need to stay alert towards the complexities of placemaking.

Indeed, we can see through the descriptions offered by the architects that this is one of their foremost concerns. BIG state that their MÉCA aims to "giv[e] Bordeaux the gift of art-filled public space from the waterfront to the city's new urban room"; Stewart Hollenstein Architects state that they opt for a largely underground library in order to

人之间的联系。最后，这些建筑是否具有公共性在于它们是否成功地促进了集会。为了不偏离相关性，这些建筑的设计者需要对空间营造的复杂性保持警醒。

实际上，我们可以通过建筑师提供的描述看出，这是他们最关心的问题之一。BIG说，他们的MÉCA文化中心项目旨在"给波尔多带来充满艺术气息的公共空间，即从海滨到城市的新城市空间"。斯图尔特·霍伦斯坦恩建筑师事务所说，他们选择了一个大型地下图书馆，以便"在地面上为户外活动留出空间，保留场地内的远景视角，为公共活动留出空间"，以及"设置一系列高度清晰的形式来创建社交空间供社区活动使用"。Paulo Mendes de la Rocha & MMBB建筑事务所的建筑师说，他们的项目通过"对现有设施进行简单的使用和改造，反映城市改造的缓慢适应过程"，最终目标是"有效地为城市如此有特色的地区实现理想复原"。Mecanoo建筑事务所和埃切西亚设计小组表示，他们设计的高雄艺术中心"是邻近的亚热带公园的一个组成部分，（为了）对人口近300万的高雄居民产生积极的社会影响"。Smiljan Radic、Eduardo Castillo和Gabriela Medrano表示，他们的设计宗旨是把比奥比奥地区剧院建成一个"伟大的艺术家及其创作作品将有机会通过自身特色丰富本地区、国家和国际文化"的地方，并使其"成为与比奥比奥河前涌现的不同建筑语言融为一体，与该地区首府市民区进行对话"的地方。

总而言之，我们在这里展示的建筑案例倾向于通过"空间营造"这个术语来定义"公共"的概念，这些是所有机构建筑中最基础的一面；除此之外，它们矗立在那儿，是作为一种邀约，同时也是一个平台，使被它们吸引而来的人们之间产生有意义的联系。但从古代开始，人们就赋予了"公共"这样的定义：它是人们集合的地方，是一个公共空间，是展示空间；通过这样的空间，每个人都可以实现自我，不是一个私人个体，而是一个社区统一体。在古代，简单的一片树荫或者一块显眼的岩石足矣；但在如今复杂的城市环境中，实现这样的公共空间可能需要更多。

"make space for outdoor activities at ground level, retaining vistas through the site and leaving space for public events", as well as "place a series of highly legible forms (…) to create social spaces for community-focused activities". Paulo Mendes da la Rocha & MMBB Arquitectos state that their project reflects the slow and adapting process of transformation of the city by "the simple use and adaptation of the [existing] facilities to (…) those that are being proposed", with the ulterior goal to "contribute effectively to the desired recovery of such a remarkable area of the city". Mecanoo architecten & the Archasia Design Group state that they designed Kaohsiung Center for the Arts "as an integral part of the adjacent subtropical park [in order] to have a positive social impact on the residents of Kaohsiung whose population counts almost 3 million". And Smiljan Radic, Eduardo Castillo, and Gabriela Medrano state that the Biobío Theater was designed as a place where "great artists and their creations, (…) will have the opportunity to enrich with identity the cultural circuit in the Region, the country and abroad" and "allow the convergence of different expressions that rise in front of the Biobío River, in dialogue with the Civic District of the regional capital".

After all is said and done, the buildings we're presenting here tend to define the notion of "public" by the very interpretation of the term as placemaking. These are least of all institutional buildings; instead they stand as concepts that resonate equally as an invitation, as well as a platform that allows for those who are drawn to them to form meaningful relationships. But this has been the very definition of "public" since antiquity: a place to assemble and to constitute a common body, a demos, through which each person is allowed to accomplish themselves not as a private individual but as a community. In the ancient days, this hardly required more than a simple shade of a tree or a most prominent rock; but in today's complex urban environments, it may take just a little bit more to accomplish the feat.

1. For a general overview of this argument see Angelos Psilopoulos, "'Le geste architectural'; Symbolism and authority in the case of the Centre Beaubourg", Architecture_MPS, Vol. 13, no. 1, February 2018.
2. Aldo Rossi, The architecture of the city, (MIT Press: Cambridge, Mass, 1982), p.33.
3. Ignasi de Solà-Morales, Differences: Topographies of Contemporary Architecture, (The MIT Press: Cambridge, Massachusetts, 1997), p.108.
4. Namely "strength, utility, grace", as translated by Granger. Vitruvius, On architecture, 2 vols, (Harvard University Press ; W. Heinemann: Cambridge, Mass; London, Vol. 1, 2015), p.35 Book.I, ch. 3, para. 2.
5. Including raising new patrons. Carl Grodach, "Cultural institutions; The role of urban design", in Tridib Banerjee and Anastasia Loukaitou-Sideris (eds.), Companion to Urban design, (Routledge: London ; New York, 2011), pp.406–407.
6. Grodach, "Cultural institutions; The role of urban design"; Aseem Inam, Designing urban transformation, (Routledge, Taylor & Francis Group: New York, 2014).
7. Phil Alden Robinson, Field of Dreams, (Universal Pictures, April 1989).
8. Aseem Inam, "Meaningful Urban Design: Teleological/Catalytic/Relevant", Journal of Urban Design, Vol. 7, no. 1, February 2002, p.35.

绿色广场图书馆和广场
Green Square Library and Plaza

Studio Hollenstein

公共建筑还是公共空间？一个全新的悉尼地下图书馆使二者兼得
Public building or public space? A new underground Sydney library provides both

这个具有前瞻性的城市项目是交付给悉尼市政委员会使用的，项目包括一个地下主图书馆和一系列如同晶莹剔透的珠宝、雕塑般矗立的地上建筑。项目还包括一个地面上的公共广场，位于绿色广场新区，这里将成为澳大利亚最大的城市改造新区，成为该地区60 500名新居民的焦点。

斯图尔特·霍伦斯坦恩建筑师事务所的设计获得了包括格伦·默卡特在内的评审团的一致通过。该设计理念作为一种独特的处理方式，融合了公共广场和图书馆这两种场所类别。这一大型地下图书馆项目有选择性地把一些建筑元素和功能生动地呈现在上层空间，地面一层为户外活动创造了空间，整个项目场地的远景得以保留，并为公共活动留出足够的空间。该项目战略性地使用了一系列易于识别的形状：圆形、三角形、长方形和梯形，为社区活动开辟出交流空间。建筑师有意识地设计了一些随意的、非正式的几何形状，营造了一种共享空间的感觉。该设计强调适应性，既要适应图书馆的不断演变发展，又要适应不断增多的人流量。

该设计对传统的图书馆建设概念提出了挑战，将其视为一系列用于掌握知识和学习的空间。在内部，图书馆围绕中央下沉的圆形花园大厅而组织设计，这个花园从上到下贯穿整个体量。花园四周有围挡，种植着茂盛的植物，儿童阅读圈还有一棵"故事树"。39个圆形天窗为地下图书馆的大厅提供经过过滤的自然光和新鲜空气。

楔形的玻璃亭子是高大而透明的入口大厅，为图书馆提供了活跃的公共交流界面，同时这里设有咖啡馆、报纸和杂志区、问讯处和自助结账机。这个入口处独立于主图书馆而运作，对公众开放的时间更长。

一座像灯塔一样的独自矗立的玻璃塔堆叠设置了一系列独立的功能空间。这座六层建筑包含一个两层楼高的阅览室（带有悬挂式夹层）、一个教学用计算机实验室（编码、机器人技术和3D打印）、一个"黑盒"剧院和用于练习与表演的空心地板音乐室（从混合打碟机到婴儿三角钢琴应有尽有）、一个名为"任意空间"且可以预订使用的灵活的社区空间，一个可以提供培训的社交型计算机实验室，顶层还有一个会议室。建筑的彩色照明设备安装在塔的顶部，置于一个透明的外壳框架中，非常醒目，提醒人们这是一个可持续的项目。为了应对悉尼的气候，图书馆的功能延伸到户外广场，其周围环绕着散发柠檬香味的桉树。这些树的树冠将最终协调邻近建筑的规模，可以提供阴凉，并通过树皮和树叶的变化表现出季节性特征。

一个下沉的室外剧场兼作图书馆的后入口，室外剧场两侧的墙上被涂成绿色，为户外表演提供了很好的场所。同时，为平整场地而设计的台阶上设置了大量的公共座位。广场设有内置电源插座的木长凳，供人们休息，做日光浴；广场上还有大面积的草坪和供儿童玩耍的水上乐园。

可持续性的特点包括：广场地面下安装了雨水储存系统用于灌溉，图书馆书架内安装了低能耗置换通风系统。精挑细选的耐旱植物以及通过收集地面排水进行的被动式灌溉系统，也可以减少对饮用水的需求。

事务所主任马提亚斯·霍伦斯坦恩说："图书馆作为高度民主的空间，在我们的城市中扮演着独特的角色。在这里，我们设计了一个城市客厅，供所有人享用，大家可以聚在这里一起讲故事。"

This visionary civic project, delivered for the City of Sydney Council, consists of a subterranean main library and a series of jewel-like above-ground sculptural building forms. It also provides a public plaza at ground level in the new district of Green Square, Australia's largest area of urban renewal, becoming a focal point for the prospective 60,500 new residents of the district.
Stewart Hollenstein's design was selected unanimously by a jury including Glenn Murcutt. In a unique response, the concept fuses the public plaza with the library. By creating a largely underground library – with selected building elements and functions emerging animatedly to the sky – the project makes space for outdoor activities at ground level, retaining vistas through the site and leaving space for public events. The project strategically places a series of highly legible forms – circular, triangular, rectangular and trapezium – to create social spaces for community-focused activities. The geometry of the design is intentionally loose and informal, establishing a sense of shared territory. The design encourages adaptation to accommodate both the evolution of the library and an expanding population.

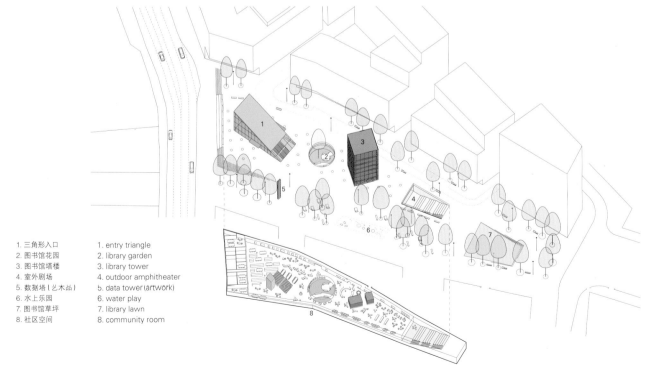

1. 三角形入口	1. entry triangle
2. 图书馆花园	2. library garden
3. 图书馆塔楼	3. library tower
4. 室外剧场	4. outdoor amphitheater
5. 数据塔(艺术品)	5. data tower (artwork)
6. 水上乐园	6. water play
7. 图书馆草坪	7. library lawn
8. 社区空间	8. community room

The design challenges the traditional concept of the library, instead conceiving it as a series of spaces for knowledge and learning. Internally the library is organized around the central sunken garden rotunda, punctured into the volume from above. This protected, lushly-planted garden incorporates a "story tree" for a children's reading circle. Thirty-nine circular skylights deliver filtered natural light and fresh air to the underground library hall.

A wedge-shaped glazed pavilion announces a lofty and transparent entrance hall and provides an active public interface for the library, accommodating a cafe that spills out onto the plaza as well as news and magazine sections, information point and self-check-out machines. The entrance pavilion is designed to operate independently from the main library, with longer public hours.

A separate, beacon-like glazed tower stacks a series of separate functions. Over six stories the tower floorplates incorporate a double-height reading room (with suspended mezzanine level), a computer lab for teaching (coding, robotics and 3D printing), a "black box" theater and sprung floor music room for practice and performance (from mixing decks to baby grand piano), and a flexible, bookable community space called the "Anything Room". A sociable computer lab offers training, and there is a boardroom space on top. The building's

一层 ground floor

地下一层 first floor below ground

1. 员工室　　　7. 接待处　　　13. 咖啡厅
2. 会议室　　　8. 图书馆花园　14. 入口
3. 公共浴室　　9. 儿童区　　　15. 共享区域
4. 图书馆藏书室 10. 图书馆休息室 16. 图书馆塔楼
5. 理事会服务中心 11. 室外剧场　17. 水上乐园
6. 阅读休息室　12. 设备室　　　18. 图书馆草坪

1. staff room
2. meeting room
3. shared bathrooms
4. library collection
5. council services center
6. reading lounge
7. reception
8. library garden
9. children's area
10. library lounge
11. outdoor amphitheater
12. plant room
13. cafe
14. entry
15. shared zone
16. library tower
17. water play
18. library lawn

colorfully-lit plant is perched on top of the tower in a transparent enclosure – a highly visible reminder of the project's sustainable focus. In response to the Sydney climate, library functions overflow into an outdoor plaza ringed by lemon-scented eucalyptus trees, the canopies of which will eventually moderate the scale of adjacent buildings, provide shade, and create seasonal character though changes in bark and foliage. A sunken, external amphitheater doubles as the rear library entrance and provides a green-walled retreat for outdoor performance, while extensive public seating is incorporated into steps created by leveling the site. The plaza also features timber sun loungers with built-in power outlets, a large lawn area and a water play zone for children.

Sustainability features include a rainwater storage beneath the plaza deck for irrigation and a low-energy displacement ventilation system within the library bookshelves. The careful selection of drought-tolerant species, and implementation of passive irrigation via surface collection drains, will also reduce the demand on potable water.

Matthias Hollenstein, Practice Director said: "Libraries play a unique role in our cities as highly democratic spaces. Here we designed an urban living room for all to share and to come together to tell stories."

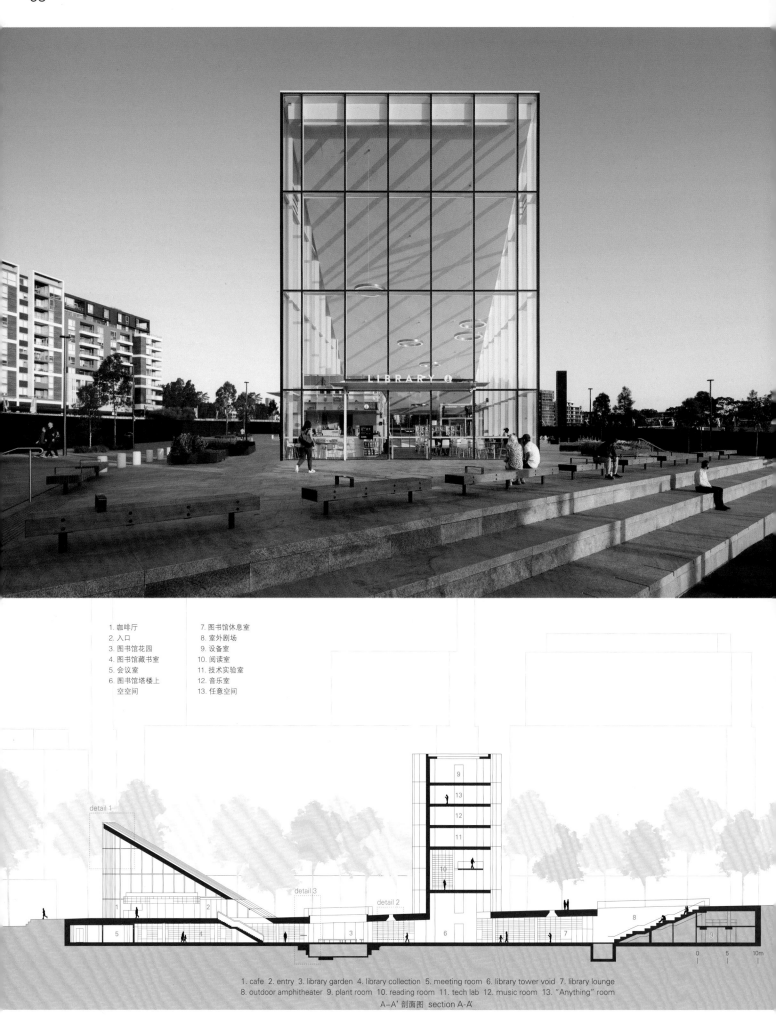

1. 咖啡厅
2. 入口
3. 图书馆花园
4. 图书馆藏书室
5. 会议室
6. 图书馆塔楼上空空间
7. 图书馆休息室
8. 室外剧场
9. 设备室
10. 阅读室
11. 技术实验室
12. 音乐室
13. 任意空间

1. cafe 2. entry 3. library garden 4. library collection 5. meeting room 6. library tower void 7. library lounge
8. outdoor amphitheater 9. plant room 10. reading room 11. tech lab 12. music room 13. "Anything" room
A-A' 剖面图 section A-A'

1. 图书馆花园 2. 共享区域 1. library garden 2. shared zone
B-B' 剖面图 section B-B'

项目名称：Green Square Library and Plaza
地点：Green Square, Sydney, NSW, Australia
事务所：Studio Hollenstein in association
with Stewart Architecture
主创建筑师：Matthias Hollenstein, Felicity Stewart
项目团队：Matthias Hollenstein, Felicity Stewart,
Matt Hunter, Simon David, Tom Vandenberg, Troy Cook,
Rachel Wan, Manus Leung, Chris Thorpe
合作者：ARUP, Hassell, Collider
顾问：landscape–Hassell; engineer–ARUP;
signage + wayfinding–Collider; public art curator–Jess Scully;
Artists of High Water in the Data Tower–Indigo Hanlee and Michael
Thomas Hill of Lightwell;
Artist of Cloud Nation suspended in library–Claire Healy and Sean
Cordeiro;
library consultant–Erik Boekesteijn
客户：City of Sydney
建筑面积：library–2,800m²; plaza–8,000m²
成本：AUD $61 million
项目开始时间：2012
竣工时间：2018
摄影师：©Tom Roe (courtesy of the architect)-p.65, p.67, p.68,
p.70, p.72, p.73; ©Julien Lanoo (courtesy of the architect)-p.62~63,
p.69

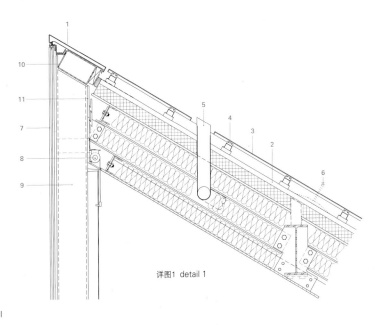

详图1 detail 1

1. parapet capping, solid aluminum pressing powder coat, white
2. 135mm Insulated panel
3. rainscreen cladding, cement composite panel
4. rainscreen cladding bracket support fixed to insulated roof panel system
5. vent pipe
6. fall arrest safety anchor
7. double glazed unit to aluminum profiles on steel frame,
   Low-e coating & heat-strengthened
8. insulated motorized blind system, automated sun sensor system,
   metalized semi-transparent blind fabric
9. 230 x 75 x 26 PFC column
10. 250 x 150 x 6 RHS steel roof beam
11. plasterboard finish

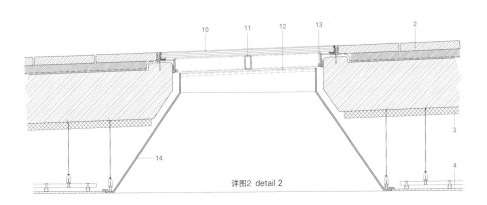

详图2 detail 2

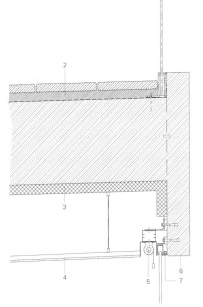

1. Balustrade:
   60mm x 40mm stainless steel top channel with toughened laminated glass cantilevered
   from stainless steel angle fixed the slab & hob
2. plaza build-up:
   - 60mm granite paver
   - 51mm grout layer
   - 6mm drainage layer with geotextile layer over, membrane protection layer under
   - 7mm waterproof membrane
3. 70mm kingspan rigid soffit insulation
4. 12.5mm perforated plasterboard, 15mm @30mm CRS round
5. insulated motorized blind system, automated sun sensor system, metalized semi-transparent blind fabric
6. drip groove
7. strip lighting
8. 30mm thick x140mm wide recycled tallowwood timber decking on steel support brackets
9. floor build-up:
   - 2.5mm Marmoleum Linoleum sheeting
   - 2.0mm Corkment underlay
   - 114.5mm topping slab
10. laminated glass skylight with sacrificial top layer & anti-slip finish
11. 50 x 100 x 6 RHS cross-brace frame
12. ø1000 circular metal framed fabric diffuser
13. light fixture
14. prefabricated plasterboard skylight cone cut parallel to match fall & direction of skylight above

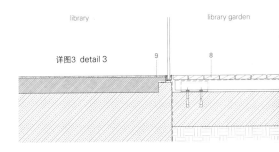

详图3 detail 3

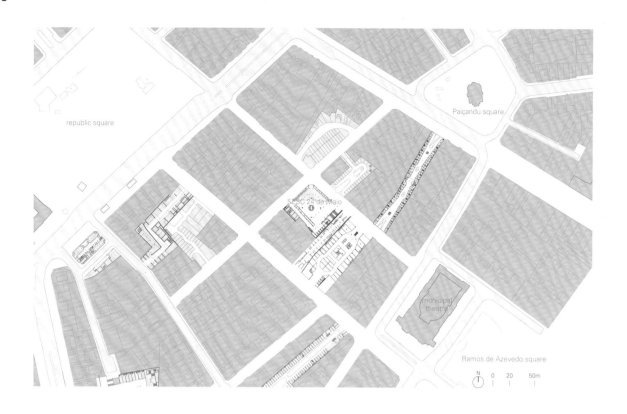

新的 de Maio 街 24 号 SESC 综合体项目为圣保罗市民提供多种娱乐和服务设施
The new SESC 24 de Maio provides various recreational and service facilities for São Paulo citizens

人们普遍认为,像圣保罗这样的城市,其转型和发展的过程是通过缓慢地适应其所处社会的习俗和生活方式的变化而实现的。新的de Maio街24号SESC综合体项目是一个集娱乐和服务设施于一身的综合项目,占据了圣保罗市中心原来的Mesbla百货商店建筑,是对介入城市文化遗址的挑战。

该设计将有效地促进城市中特色区域的升级,并对建筑进行重新调整以满足全新的用途和特定的项目需求。为此,建筑师提出了下列基本构想:

1.把一层可以容纳一个大型开放广场的空间作为公共场所建设的中心,将其设计成一条具有画廊风格的自由通道,并直接连接到活跃的社区中;

2.将原来的地下车库改造成咖啡厅和剧院,与热闹的街道建立明确的联系;

3.创建一个新的垂直交通系统,该系统由一系列宽敞的坡道组成,勾勒出一个清晰连续的回路,设置一条建筑长廊,并且巧妙地将城市更广阔的空间与SESC新建筑的各种室内项目联系起来;

4.从战略层面考虑,构建各种开放空间,将其设计为有顶的架高广场。这些空间包括娱乐广场和游泳池花园,没有全部用立面封闭,而是被设计为一个沿着开放的循环网状坡道延伸的空中花园;

5.为了不使简单重叠型楼层显得单调,最终采用双层大厅和双层立面与架空走廊联合使用的建设方案;

6.在屋顶上建造了一个宽敞的广场,一个拥有开放游泳池的日光浴室;

7.将所有技术和机械基础设施集中在一个附属结构中,把它作为一个独立的服务塔,建在一座多年前就被废弃的烂尾楼上;

8.将一套家具融入建筑设计中。这启发了一系列丰富多彩的设计灵感,包括椅子、桌子、沙发、课桌、柜台和模块化堆叠的家具元素,以愉悦有趣的方式完成了建设并彰显了其机构特征。

这些目标是设计中挑战性任务的核心,明确了空间特色的建造方向。为了实现这些目标,建筑师就建筑技术和基础设施的供应做出以下决定:

1.有选择地拆除原有建筑结构的特定部分,同时又保持中央大厅和基本结构构件不变,创造垂直的空间;

2.建造独立的新建筑结构,由四根主要柱子支撑,这四根柱子组成中央空间,留下一系列大型交错排列的大厅及屋顶上带有游泳池的日光浴室;

3.对原有车库地下区域进行稍微下沉处理,以满足剧院及其附属体量的要求,充分尊重邻近建筑的限制条件和结构基础;

4.开发一套覆盖建筑所有楼层的复杂的烟雾应急控制机械系统,它的复杂性支持双层楼板和开放式通道网格这样的空间安排。

The process of transformation and development of cities like São Paulo is believed to be made by slowly adapting to the changes of customs and way of life of the societies that build them. The new SESC 24 de Maio, a complex of recreational and service facilities, occupies the old Mesbla department store in downtown São Paulo, and represents a challenge of intervening in the context of an urban heritage site.

The design is set to contribute effectively in upgrading such a remarkable area of the city and to readapt the building structure to a completely new set of uses and specific programs. In order to do so, the following basic ideas were outlined:

1. To house a large open square at ground level as a central idea of public placemaking, designed as a free passage with a gallery character that directly connects to the exciting neighborhood;

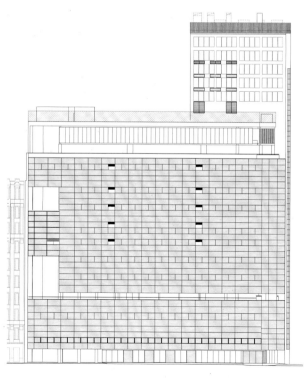

东北立面 north-east elevation

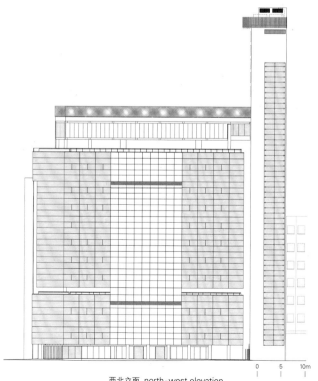

西北立面 north-west elevation

2. To transform the former underground garage into a cafe and a theater, in explicit relation to lively streets;

3. To create a new vertical circulation system composed of a generous sequence of ramps, outlining a clear and continuous circuit that sets an architectural promenade and playfully connects the wider urban spaces of the city to the various indoors programs of the new SESC building;

4. To structure a variety of open spaces at strategic levels that are designed as covered elevated squares. Such spaces, including conviviality square and the swimming pool garden, are not enclosed with facades and act as hanging gardens along the open circulation network of ramps;

5. To adopt eventual associations of a great hall of two levels and double-skin facade, with the overhead galleries, in order to avoid the monotony of simple overlapping type floors;

6. To build a generous square on the rooftop, a solarium that includes an open swimming pool;

7. To concentrate all technical and mechanical infrastructure on an annex structure that acts as an isolated service tower, built in a contiguous property that had been abandoned years before;

8. To incorporate a set of furniture pieces to the architectural design. This has resulted in a colorful design series of chairs, tables, sofas, desks, counters and modular stackable furniture elements that complete the building in a joyful and playful manner, strengthening its institutional identity.

To bring about such goals, which are at the core of the design challenges and which orient its spatial character, decisions on building techniques and infrastructural supply were adopted:

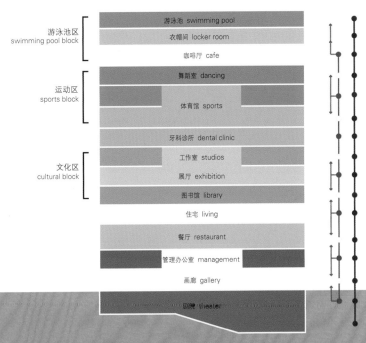

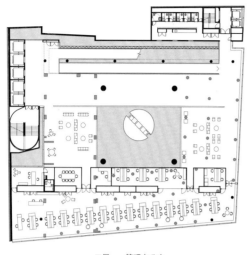

二层——管理办公室
first floor _ management

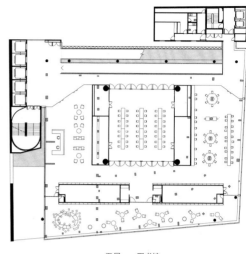

五层——图书馆
fourth floor _ library

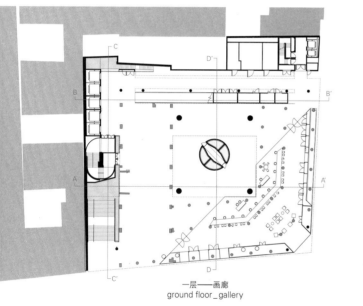

一层——画廊
ground floor _ gallery

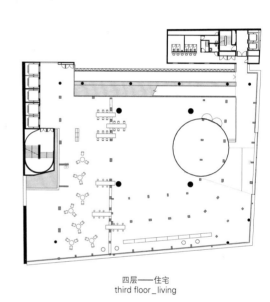

四层——住宅
third floor _ living

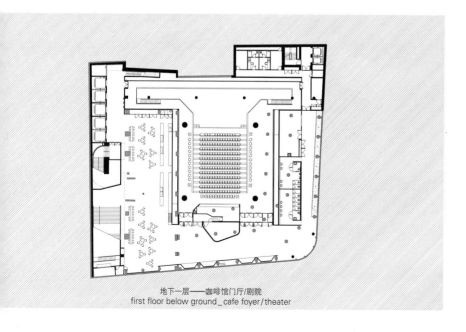

地下一层——咖啡馆门厅/剧院
first floor below ground _ cafe foyer / theater

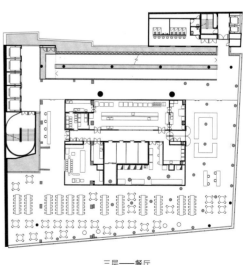

三层——餐厅
second floor _ restaurant

83

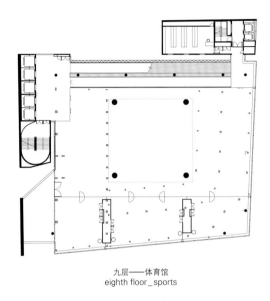

九层——体育馆
eighth floor _ sports

屋顶——游泳池
roof _ swimming pool

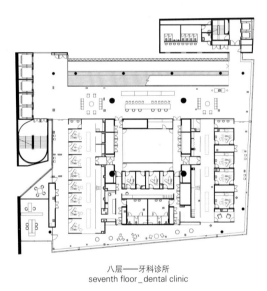

八层——牙科诊所
seventh floor _ dental clinic

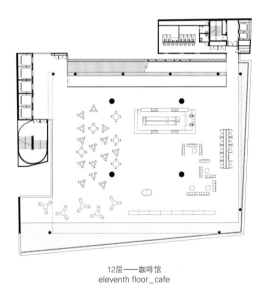

12层——咖啡馆
eleventh floor _ cafe

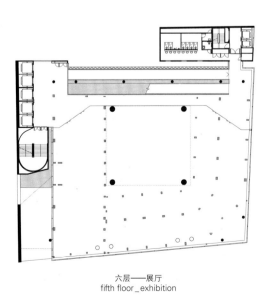

六层——展厅
fifth floor _ exhibition

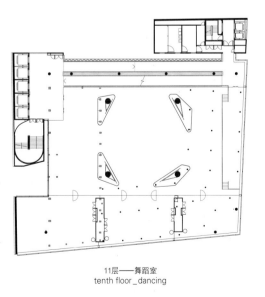

11层——舞蹈室
tenth floor _ dancing

1. The selective demolition of specific parts of the original building structure, while maintaining its basic structural elements and central hall, creating the vertical void;
2. The construction of a new and autonomous building structure, supported by four main pillars that configure the central void, sustaining a series of large interleaved halls and the solarium with a swimming pool on the rooftop;
3. The adaptation of the building's former garage to a theater by slightly lowering its level while carefully respecting the limits and structural foundations of its neighboring building;
4. The development of an elaborate mechanical emergency system for the smoke exhaustion at all levels. Its complexity allowed spatial arrangements, such as double floors and an open circulation network.

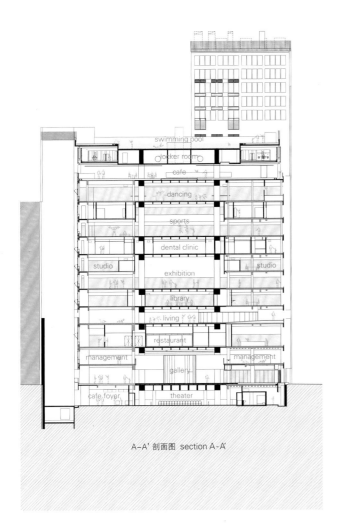

A-A' 剖面图 section A-A'

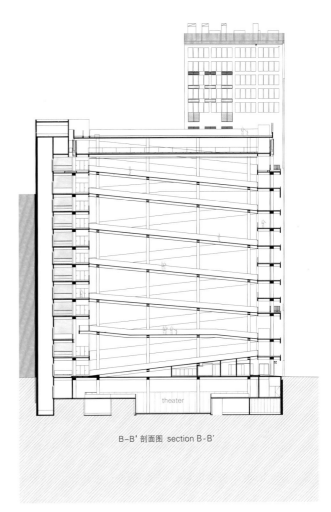

B-B' 剖面图 section B-B'

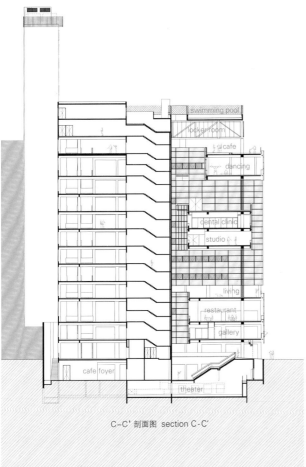

C-C' 剖面图 section C-C'

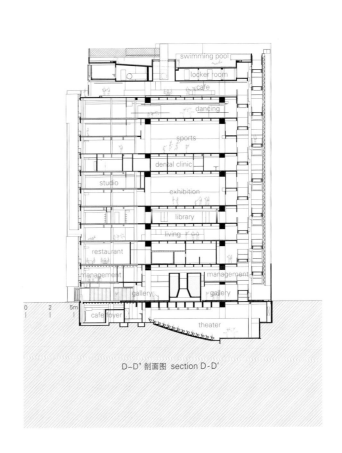

D-D' 剖面图 section D-D'

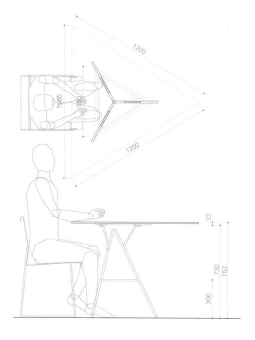
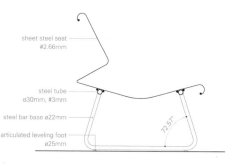

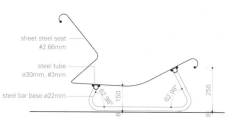

项目名称：SESC 24 de Maio / 地点：24 de Maio Street X Dom José de Barros Street, Downtown, São Paulo, SP, Brasil / 事务所：Paulo Mendes da Rocha + MMBB Arquitetos / 项目员工：Fernando Mello Franco, Marta Moreira, Milton Braga, Adriano Bergemann, Ana Carina Costa, Ana Carolina Mamede, Cecília Góes, Eduardo Ferroni, Giovanni Meirelles, Gleuson Pinheiro, Guilherme Pianca, Jacques Rordorf, Lucas Vieira, Márcia Terazaki, Maria João Figueiredo, Marina Acayaba, Marina Sabino, Martin Benavidez, Vito Macchione, João Yamamoto, Rafael Monteiro, Rodrigo Brancher, Thiago Rolemberg, Victor Olivera / 建筑结构：Kurkdjian e Fruchtengarten Engenheiros Associados / 建筑基础：MAG Projesolos Engenheiros Associados / 电气与液压项目：PHE Projetos Hidráulicos e Elétricos 施工：Mendes Júnior Trading e Engenharia S/A / 工程员工：Eng. Amilcar João Gay Filho, Eng. Humberto Bigaton, Eng. Alberto Costa Souza Neto / 建筑面积：27,865m² / 设计时间：2002—2008 / 施工时间：2012—2017 / 竣工时间：2017 / 摄影师：©Nelson Kon

MÉCA 文化中心
MÉCA Cultural Center
BIG

BIG 为波尔多市的城市生活创建了一个重要的新文化中心
BIG creates a major new cultural hub for the urban life of Bordeaux

MÉCA文化中心全称为"Maison de l'Économie Créative et de la Culture en Aquitaine",面积18 000m², 为举行当代艺术、电影、表演的活动提供了场所,为波尔多市带来了一个从海滨漫延到新城市空间的充满艺术气息的公共空间。MÉCA位于Garonne河和Saint Jean火车站之间, 将三个地区艺术机构——展现当代艺术的FRAC, 展现电影、文学和视听艺术的ALCA以及展现表演艺术的OARA——连接在一起,巩固了这个联合国教科文组织遗产保护城市作为文化中心的地位。

该建筑被设计成一个连接文化机构和公共空间的拱形结构。人行道被延长,成为通向城市客厅空间的通道。通过外立面,可以瞥见OARA的舞台塔、ALCA的办公室和FRAC沐浴在透过屋顶天窗洒进室内的日光下的艺术画廊。一系列台阶和坡道直接引导公众进入MÉCA核心地带1100m²的户外城市空间,创造了一座多孔的机构建筑,参观者可以通过人行道体验在Quai de Paludate大街和河边长廊之间自由漫步。

一个装有白色LED灯的7m高的MÉCA标志照亮了整个空间,如同这个城市空间中颇具现代气息的吊灯。在有特殊需要的时候,可以把MÉCA的户外空间改造成音乐会或大型戏剧演出的舞台,或作为雕塑和其他艺术设施的外延陈列室。由法国艺术家Benoît Maire设计的永久性青铜雕塑——一个赫尔墨斯(希腊神话人物)半个面部的雕塑作品,坐落在河滨一侧的入口处,引导参观者思考该地区的当代文化。

参观者从一层进入MÉCA文化中心之后首先到达的是大厅,在这里他们可以在螺旋区域里面休息,也可以在Le CREM餐厅吃饭,餐厅中摆放着由BIG公司设计的红色家具和软木椅子,用以致敬这座以葡萄酒闻名的城市。参观者还可以通过餐厅与电梯旁的巨大观测镜看室外的城市活动,在室外也可以通过它看到建筑内的情况,从而使室内和室外之间形成了对话。购票的参观者可以在同一层的OARA的剧场欣赏表演。这里设有250个配置灵活的座椅,音响系统通过由混凝土、木材和

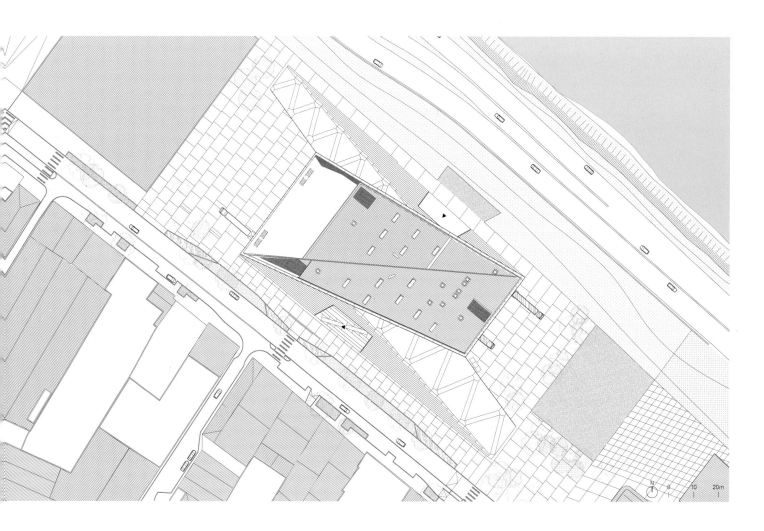

项目名称: MÉCA Culture Center / 地点: 54 Quai de Paludate, 33800 Bordeaux, France / 事务所: BIG / 合伙人负责人: Bjarke Ingels, Jakob Sand, Finn Nørkjær, Andreas Klok Pedersen Project manager: Laurent de Carnière, Marie Lancon, Gabrielle Nadeau / 项目团队: Alexander Codda, Alicia Marie Sarah Borhardt, Annette Birthe Jensen, Åsmund Skeie, Aya Fibert, Bartosz Kobylakiewicz, Bernhard Touzet, Brigitta Gulyás, David Tao, Edouard Champelle, Espen Vik, Greta Krenciute, Greta Tafel, Hyojin Lee, Ivan Genov, Jan Magasanik, Jeffrey Mark Mikolajewski, Karol Bogdan Borkowski, Katarzyna Swiderska, Kekoa Charlot, Lorenzo Boddi, Maria Teresa Fernandez Rojo, Melissa Andres, Michael Schønemann Jensen, Nicolas Millot, Ola Hariri, Ole Dau Mortensen, Pascale Julien, Paul-Antoine Lucas, Raphael Ciriani, Santiago Palacio Villa, Se Hyeon Kim, Sebastian Liszka, Seunghan Yeum, Snorre Emanuel Nash Jørgensen, Teresa Fernández, Thiago De Almeida, Thomas Jakobsen Randbøll, Yang Du, Zoltan David Kalaszi, Tore Banke, Yehezkiel Wiliardy 合作者: FREAKS freearchitects, Lafourcade-Rouquette Architectes, ALTO Ingénierie, Khephren Ingénierie, Hedont, dUCKS Scéno, Dr. Lüchinger + Meyer Bauingenieure, VPEAS, Ph.A Lumière, ABM Studio, Mryk & Moriceau, BIG Ideas / 客户: Région Nouvelle-Aquitaine / 功能: culture / 总楼面积: 18,000m² / 竣工时间: 2019
摄影师: ©Laurian Ghinitoiu (courtesy of the architect)

穿孔金属组成的全黑的格子状面板得以优化。在建筑的楼上，影迷们可以在ALCA的80座影院内观看电影，或者参观两个制作办公室和项目孵化区。

FRAC位于上部楼层，有7m层高的展览空间、艺术家制作工作室、存储设施、90个座位的礼堂和咖啡厅。屋顶露台的面积约为850m²，成为灵活的展区延伸区域。在这里可以欣赏整个城市和圣米歇尔大教堂的景色，未来这里可以举办大型户外艺术展览和演出。MÉCA文化中心的建筑立面几乎完全由4800块预制混凝土板建造而成，镶嵌着尺寸不一的窗户，以控制进入建筑内部的日光量，从而营造一种通透感。混凝土板材的重量有的重达1.6t，经过打磨，展现出原始的石材质感，其纹理如同波尔多当地砂岩的表面纹理。明亮而温暖的黄色沙砾在阳光下辐射着建筑，使MÉCA文化中心与当地城市乡土文化融合，并成为一个崭新却似曾相识的景观。

The new 18,000m² Maison de l'Économie Créative et de la Culture en Aquitaine, MÉCA, creates a framework for the celebration of contemporary art, film and performances, giving Bordeaux the gift of art-filled public space from the waterfront into the city's new urban room. Centrally located between the River Garonne and Saint Jean train station, MÉCA loops together three regional arts agencies – FRAC for contemporary art, ALCA for cinema, literature and audiovisuals, and OARA for performing arts – cementing the UNESCO-listed city as an epicenter of culture.

The building is conceived as an arc of cultural institutions and public space. By extruding the pavement of the promenade, this becomes the ramp that leads into the urban living room;

同一座建筑中有三个机构
three institutions in one building
How do you preserve the autonomy of three independent cultural institutions while still allowing for maximum interdisciplinary exchange and accessibility to the public? We propose merging the FRAC, ALCA and OARA into a single loop framed around a fourth program: an outdoor urban room.

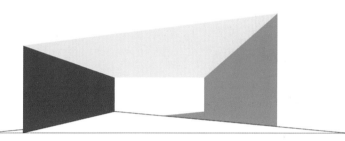

连续的步行道
continuous promenade
The outdoor urban room invites the arts and everyday life of Bordeaux to flow from the city center to the Garonne riverfront. The pavement of the promenade is extruded to become the ramp that leads into the urban room, the facade with glimpses into the OARA and ALCA, and the rooftop enclosing the FRAC galleries. During special occasions, MÉCA's outdoor spaces can transform into a stage for concerts and theatrical spectacles or a gallery for art installations.

场地上的基础体量
basic volume on site

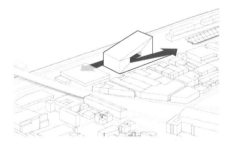

线性步行道的连续性
continuity of the linear promenade

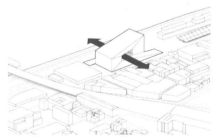

穿过建筑的公共通道
public passage through the building

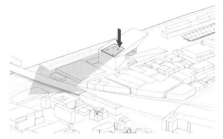

俯瞰波尔多的全景屋顶露台
panoramic roof terrace overlooking Bordeaux

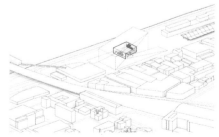

向公众开放的艺术品城市空间
art-filled urban room for the public

作为文化中心的人居景观
inhabitable landscape as the epicenter for culture

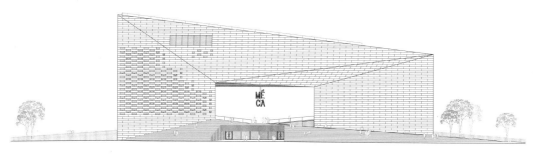

北立面 north elevation

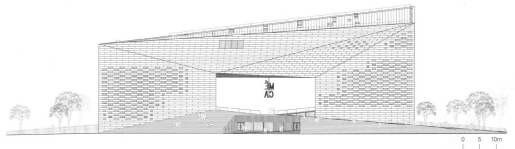

南立面 south elevation

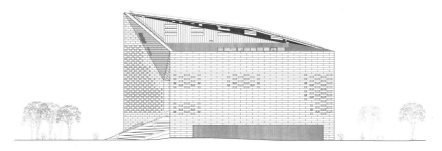

西立面 west elevation

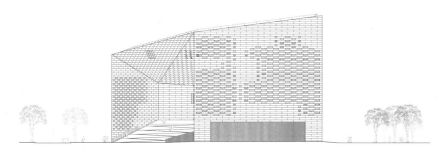

东立面 east elevation

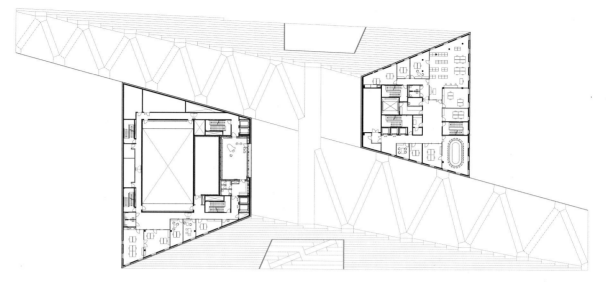

三层 second floor

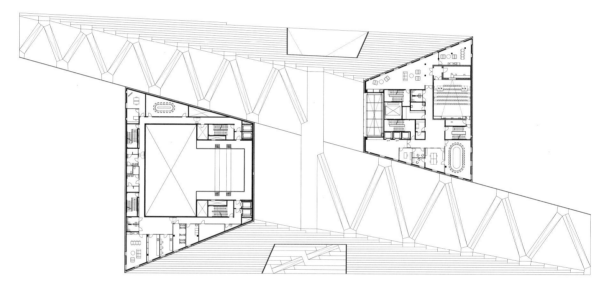

二层 first floor

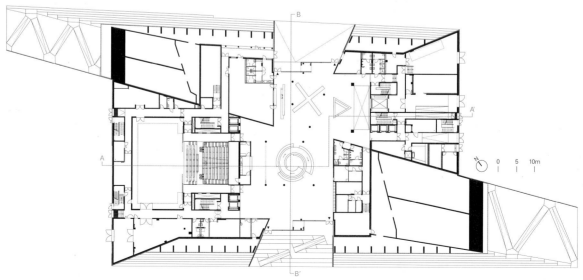

一层 ground floor

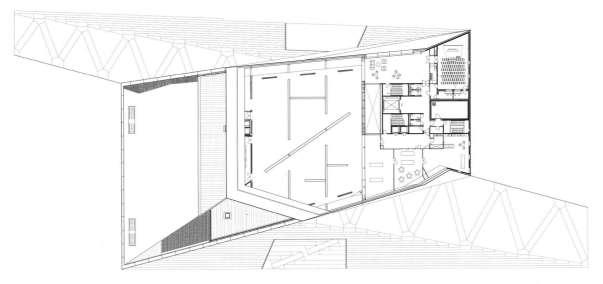
七层 sixth floor

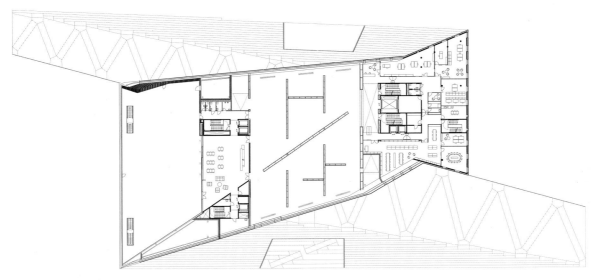
六层 fifth floor

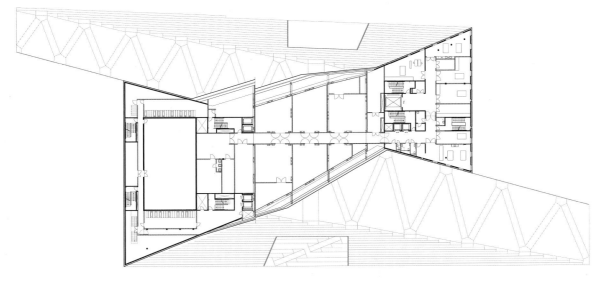
五层 fourth floor

the facade offers glimpses into the stage towers of OARA and the offices of ALCA, and the rooftop encloses the sky-lit galleries of FRAC. A series of steps and ramps lead the public directly into the 1,100 m² outdoor urban room at the core of MÉCA, creating a porous institution for visitors to roam freely between the Quai de Paludate street to the river promenade. A 7m-high MÉCA sign illuminates the space with white LED lights, like a modern chandelier at the scale of the urban room. On special occasions, MÉCA's outdoor spaces can be transformed into a stage for concerts and theatrical spectacles or an extended gallery for sculptures and other art installations. A permanent bronze sculpture depicting a half-head of Hermes by French artist Benoît Maire intersects with the entrance on the riverside, inviting visitors to reflect on the contemporary culture of the region.

Upon entering MÉCA from the ground floor, visitors arrive at the lobby where they can relax in the spiral pit or dine at the restaurant Le CREM, furnished with red furniture and cork chairs designed by BIG in reference to the city known for its wine. A giant periscope by the restaurant and elevators allows visitors to see the activity in the outdoor urban room and vice versa, creating an indoor-outdoor dialogue. On the same ground floor, those with tickets can enjoy performances in OARA's 250-seat theater featuring flexible seating configurations and acoustic systems optimized by an all-black checkerboard panel of concrete, wood and perforated metal. Upstairs, filmgoers can view screenings at ALCA's red-accented 80-seat cinema or visit the two production offices and project incubation area.

FRAC occupies the upper floors with 7m-high exhibition spaces, production studios for artists, storage facilities, 90-seat auditorium and cafe. The 850 m² public roof terrace serves as a flexible extension to the exhibition spaces, allowing future large-scale art installations and performances to be held outdoors amid views of the city and the Basilica of St Michael. MÉCA's facade is composed almost entirely of 4,800 prefabricated concrete panels interspersed with windows of various sizes to control the amount of light entering inside and to create a sense of transparency. The concrete slabs, which weigh up to 1.6 tons, are sandblasted to expose its raw qualities and to texture the surface with the local sandstone of Bordeaux. Yellow granules for brightness and warmth radiate the building in the sun and help to integrate MÉCA with the city vernacular as a new yet familiar sight.

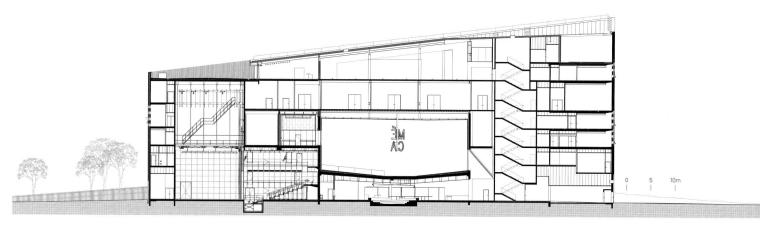

A-A' 剖面图 section A-A'

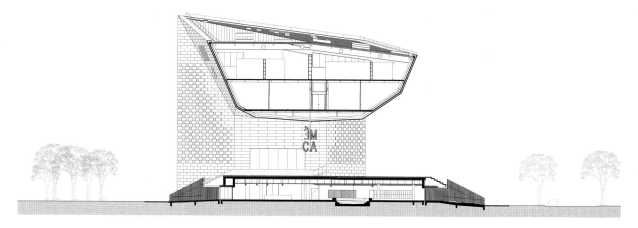

B-B' 剖面图 section B-B'

比奥比奥地区剧院
Biobío Regional Theater

Smiljan Radic + Eduardo Castillo + Gabriela Medrano

比奥比奥地区剧院半透明的骨架结构使剧院本身成为风景
Biobío Regional Theater's translucent and skeletal structure creates scenery of the theater itself

我把它藏起来,封起来,我的包装就是试图"感知"事物的本质。
——塔德乌什·康托尔,Wielopole/Wielopole

尽管比奥比奥地区剧院距离2010年智利发生的里氏8.8级地震的震中不是很远,而且该地区在地震中也受到波及,但剧院的基础并不是很深。

整座建筑建在30cm厚的混凝土板上,四周由50cm厚的加固梁支撑。地面一层的对角支撑形成了一个刚性的表面隔膜,与二层和基础板一起,使整座建筑看起来似乎在飘移,不想在地面上留下任何足迹。这种奇怪的感觉,加上半透明聚四氟乙烯的外层包裹,使它在某些方面与马戏团的帐篷有几分相似。观看该剧院的结构截面,或它的基础平面,是一件让人感到愉悦的事。

建筑师的设计理念是该建筑就是一个被包裹的骨架。当观众在剧院里面走动的时候,如同在各个框架和支撑中穿梭。这样的网格给人的印象是每个空间角落的大小尺寸都是相互制约的。所有这些围绕着房间的结构设备都暗示着这里是中间空间,而房间本身,就像结构部分中间的开放场地。整个结构就是一个脚手架,是幕布的反面,舞台布景的下层支撑结构,这些通常都是隐藏的,看不见的。

这样一来,观众们不必等到穿过前厅进入黑暗的主厅才能揭开剧院神秘的面纱。在人们进入剧院之前就感觉到它的神秘了。来访者只需要看到裹在建筑表面的"布",去感受、去感知藏在里面的东西,就像康托尔说的那样,或至少有那么一刻会相信,倘徉其中,将会是一种非凡的体验,至少是一种令人惊讶的体验。

另一方面,演员们可以接触到一种"灵活的格调"。大厅具有当代的规模,拥有一些基本的机械仪器,使室内空间在技术上达到了多功能空间必要的最佳效果。

My wrappings were an attempt to "sense" the nature of the object. Hiding it, wrapping it.
– Tadeusz Kantor, Wielopole/Wielopole

The Biobío Regional Theater does not have deep foundations, despite being located just a short distance from the epicenter of the earthquake that struck this area of Chile in 2010 with an intensity of 8.8 on the Richter scale.
The entire building is constructed on a 30cm-thick concrete slab, with 50cm-thick reinforcement beams around the perimeter. The diagonal of the ground floor forms a rigid surface diaphragm, along with the first floor and foundation slab, almost as if the building were drifting and did not want to leave tracks. This strange sensation, together with the building's outer wrapping of translucent PTFE, makes it in some respects the distant cousin of a circus tent. Seeing a structural cross section of the theater, or its foundation plans, becomes a thing of joy.

项目名称：Biobío Regional Theater / 地点：Concepcion, Bio Bio Region, Chile / 建筑师：Smiljan Radic, Eduardo Castillo, Gabriela Medrano / 结构工程师：ByB ingeniería Ltda. 客户：Ministry of Culture and Arts, Ministry of Public Works Government of Chile / 功能：main hall for 1200 persons and concert hall for 250 persons. / 用地面积：20,050m² 建筑面积：9,650m² / 结构：reinforced concrete / 材料：reinforced concrete and PTFE membrane / 设计时间：2012—2013 / 施工时间：2015—2017 / 摄影师：©Cristobal Palma

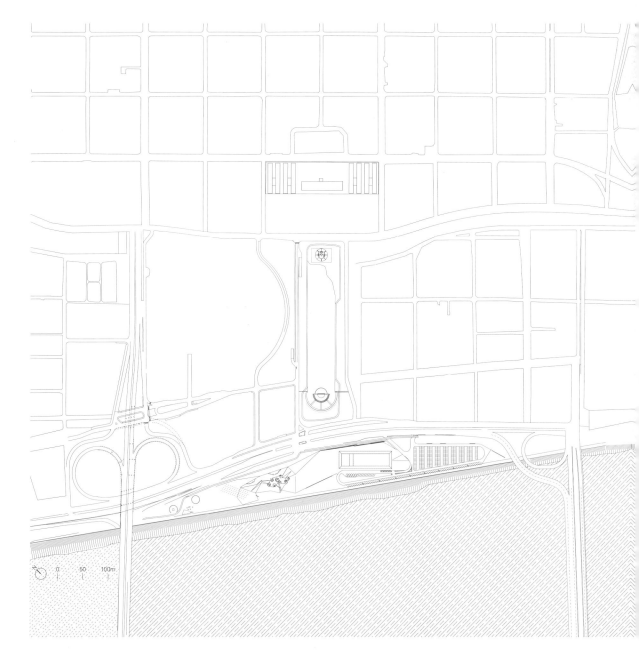

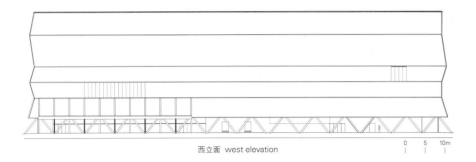
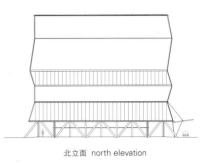

西立面 west elevation

北立面 north elevation

The architects' concept for the project is that of a wrapped skeleton. Within the theater, spectators move through a space which is punctuated continuously by struts and supports – a grid which gives the impression that each corner is being measured against another corner. All this structural paraphernalia around the rooms suggests the intermittent spaces, and the rooms themselves, are like open fields in the midst of a structural section. The structure is simply a scaffold, the reverse side of a backdrop, the inferior support structure of stage scenery that is normally hidden and unseen.

In this way, spectators do not have to wait until they have crossed the vestibule and entered the darkness of the main hall to open up the theater; the mystery appears even before entering. Visitors have only to see the "cloth" that falls over the building, veiling it, to feel or "sense", as Kantor says, that something is hidden inside – or at least to believe for a moment that moving around inside it will be an experimental process, one that is, at the very least, surprising.

Actors, on the other hand, have access to a "flexible air". The halls acquire contemporary dimensions, and with some basic mechanical instrumentation, the necessary optimum technical qualities for the multiple uses of the interior space are achieved.

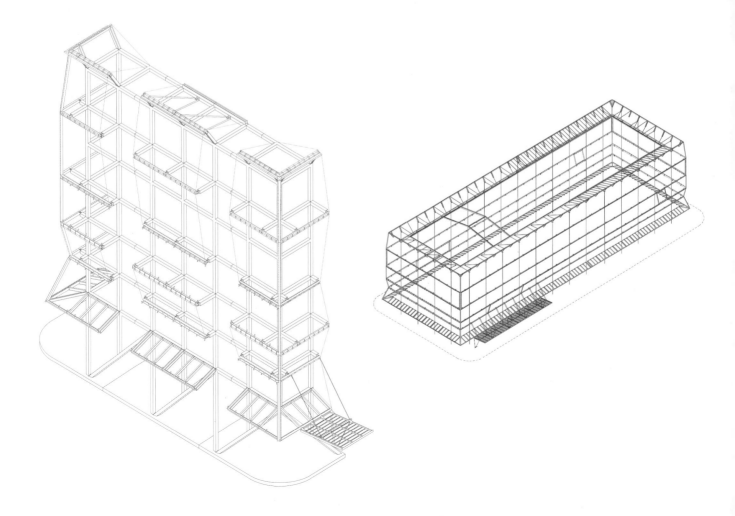

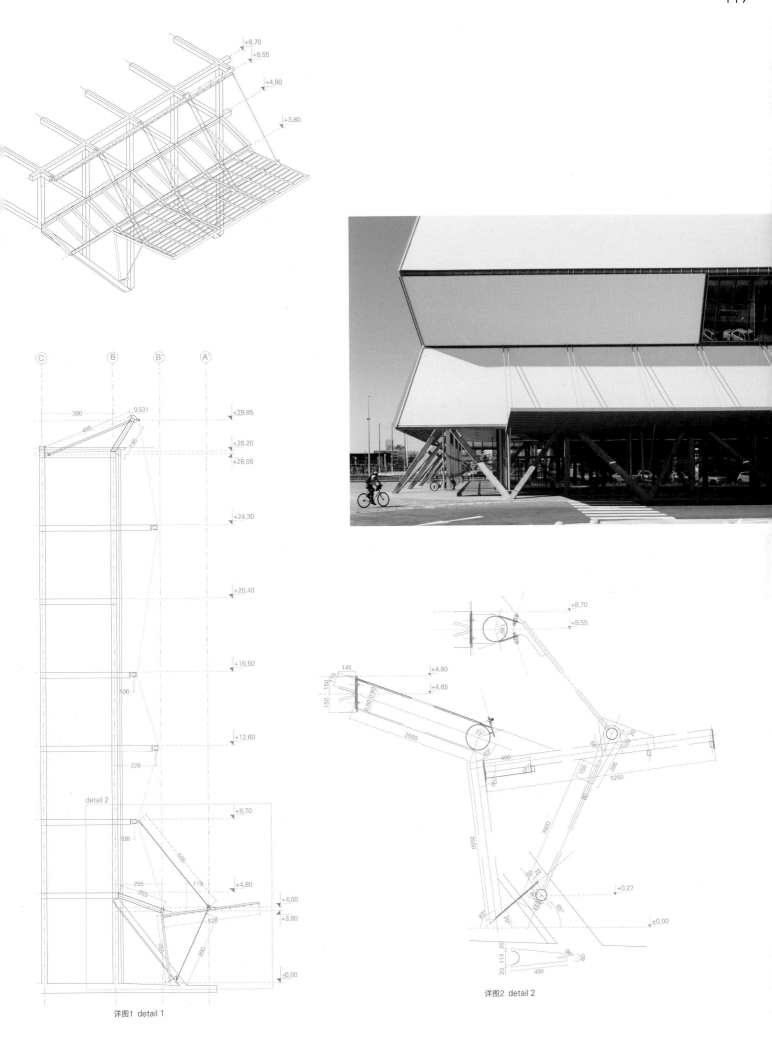

详图1 detail 1

详图2 detail 2

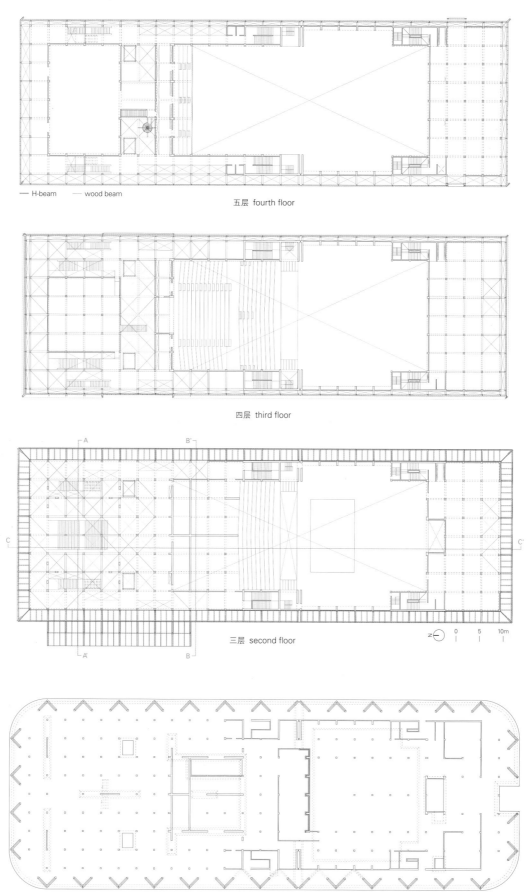

—— H-beam　—— wood beam　　五层 fourth floor

四层 third floor

三层 second floor

二层 first floor

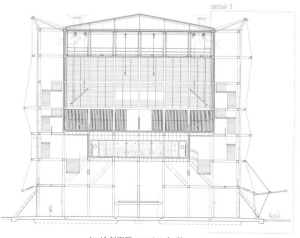

A-A' 剖面图 section A-A'

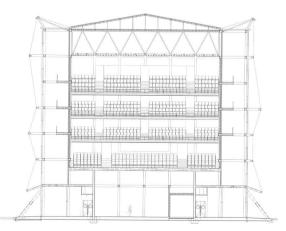

B-B' 剖面图 section B-B'

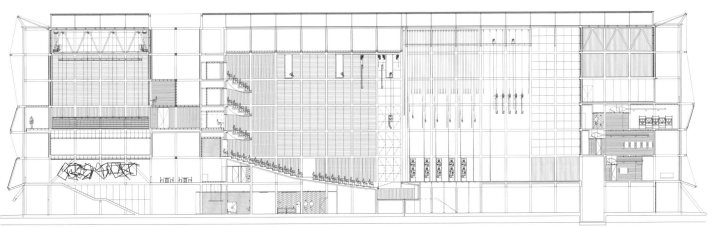

C-C' 剖面图 section C-C'

高雄艺术中心
Kaohsiung Center for the Arts
Mecanoo architecten

高雄艺术中心的设计灵感来自于榕树波浪形树冠
Kaohsiung Center for the Arts derives inspiration from the undulating canopy of the Banyan tree

高雄市拥有近三百万人口,是一个随性的、充满勃勃生机的城市。它不仅是中国台湾的第二大城市,也是世界最大海港之一,兼有亚热带台风、高温、强降雨等极端气候,地震也频发。这一面积为140 000m² 的新建综合体必须能应对所有这些极端情况。

艺术中心位于高雄市的中心地带。这里原来是一个650 000m²的军事基地,直到2006年,这里还是一个封闭的军营,里面尽是流浪狗,榕树丛生。

经过几十年的时间,这些扭曲的树枝逐渐交错在一起。向上生长的枝条垂向地面,延伸到土壤中,所覆盖的范围非常广泛,令人震惊。这些新生长的树冠连成一片,如同一个巨大的雨伞遮盖着这片区域。

榕树开放的、保护伞的形状成为设计的灵感。建筑巨大的、波浪形起伏的结构是由一家当地公司和一家荷兰的造船厂合作建成的,将各种不同的功能区域连为一体。

这个宽敞封闭的树冠结构下面是榕树广场,一个半封闭的公共空间,凉爽的微风在此自由流动。这是一个宽敞的公共空间,无论是白天还是晚上,居民们都可以在这里自由地闲逛,打打太极拳,或在人行道和非正式空间里进行街头表演。建筑的屋顶表面有一处向地面弯曲,呈曲线状,犹如一个露天剧场,而周围的公园就如同舞台。

四个正式的表演大厅之间的部分就是支撑着屋顶的"树干",它们从地面向上上升到5m以上,成为公园景观的一部分。而屋顶与地面接触的部分则形成一个露天剧场,向周围的公园敞开,同时既是舞台又是场景。

这些不同的剧院中有三个可容纳约2000人,各剧院通过位于屋顶之下的各个休息室和一个地下服务楼层连为一体。这个地下服务楼层是每个剧院的后台服务区域。

音乐厅的形状像一个呈阶梯式的葡萄园,中间是一个舞台,不同高度的观众席环绕四周,使观众离舞台更近,可以获得更佳的视听体验。观众们从一层进入这个衬有橡木衬板的礼堂,螺旋形的坡道将他们引向金色的座位。

歌剧院被设计成马蹄形,有三个半圆形的观众席,座位上有红色和紫色布料的软垫,软垫上有台湾花卉图案,与深色的墙壁形成了对比。剧院既可以上演西方歌剧,容纳超过70名音乐家组成的管弦乐队,

也可以通过操纵一个悬挂的声学装置，使其适合上演中式戏曲表演。

剧场有着和Mecanoo事务所标志色一样的蓝色的座椅，适合举办各种戏剧和舞蹈表演。灵活性是这个多功能空间设计的核心元素。

在独奏厅的设计中，表演者和观众之间有着最亲密的气氛，不对称的空间安排和上下两层大约400个座位的排布方式，主要是为室内乐演奏和欣赏而设计的。座位采用与音乐厅相同的金色面料，并有橡木衬板覆盖墙壁。独奏厅的上部围了一圈吸声幕帘，使空间内的混响时间可以根据特定类型的表演进行调节。

Kaohsiung is an informal, lively city of almost three million inhabitants. Not only is it the second largest city in Taiwan, China, and one of the world's largest sea ports, but it is also host to a dramatic sub-tropical climate of typhoons, high temperatures, heavy rainfall and regular earthquakes. The new 140,000m² performance complex must cope with all of these extremes.

The location is a former 650,000m² military compound in the center of the metropolitan area. Until 2006 the site was a fenced-off military barracks, occupied only by stray dogs and dense with banyan trees.

Over decades the twisting branches of the trees gradually interlocked, their aerial shoots grew down into the soil (spreading over an astonishingly wide area) and the crowns of these new trunks merged into one, like a huge umbrella covering the site.

The open, protective shape of the banyan trees became a springboard for the design. The vast, undulating structure, built in cooperation between a local and a Dutch shipbuilder, is composed of a skin and roof, and connects an extensive range of functions.

Underneath an expansive, sheltered crown is the Banyan Plaza, a partially enclosed public space where cooling breezes freely flow. It is a generous public space; residents can now wander

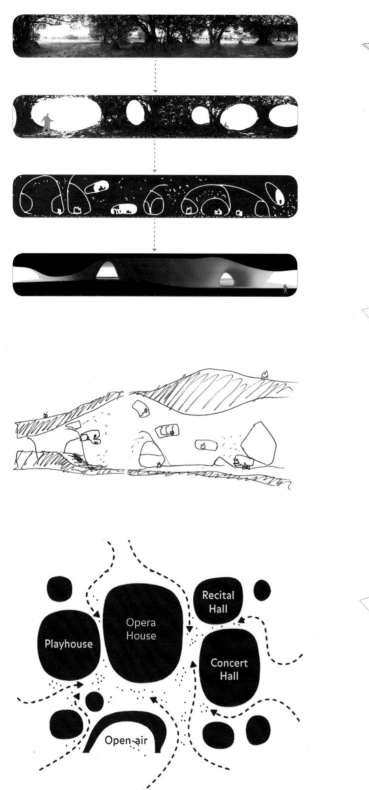
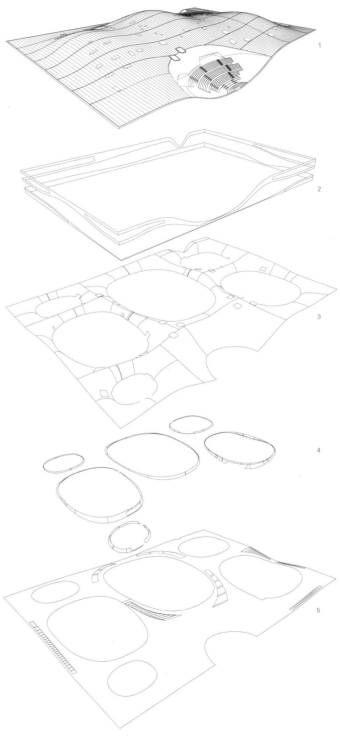

1. 屋顶设有室外座位（34 843m²） 2. 垂直立面（6896m²）
3. 榕树广场表皮（20 724m²） 4. Fillet（2320m²） 5. 榕树广场（17 446m²）
1. roof including outdoor seating (34,843m²) 2. Vertical facade (6,896m²)
3. Banyan plaza skin (20,724m²) 4. Fillet (2,320m²) 5. Banyan plaza (17,446m²)

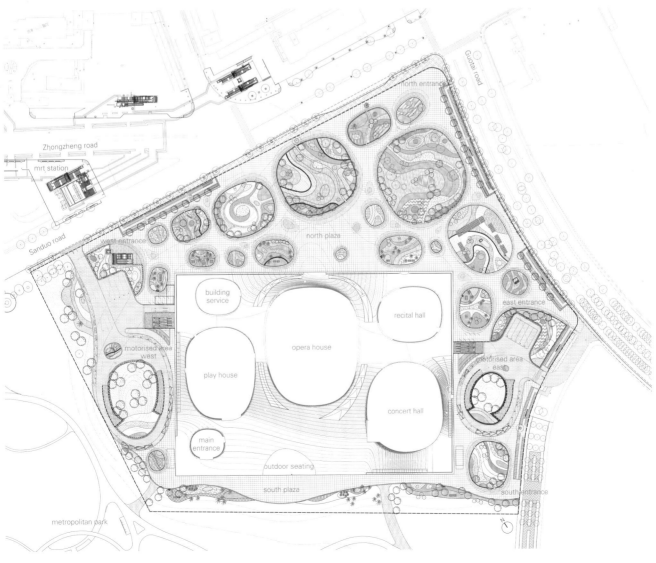

through here day and night, practice Tai Chi or stage street performances along walkways and in informal spaces. An open-air theater nestles on the roof, where the structure curves to the ground, with the surrounding park forming the stage.

Between the four formal performance halls – the "trunks" that support the roof – a topography rising from ground level to over five meters becomes part of the park's landscape. Where the roof touches the earth, the building becomes an amphitheater, open towards the surrounding parkland which, in turn, acts as both stage and set.

These different theaters, three of which have a capacity of around 2000, connect via foyers in the roof and an underground service floor which houses the backstage area of each theater.

The concert hall is shaped as a stepped vineyard with a stage at its center, such that terraces at different floor heights encircle the podium, bringing the audience in close proximity to the performance. The public enters the oak-lined auditorium from the ground floor and spiraling ramps lead them to their golden seats.

The opera house is arranged as a horseshoe with three semi-circular balconies. The seating is upholstered in red and purple fabrics with a pattern of Taiwanese flowers, contrasting with the darker walls. This theater is suitable for Western opera, with an orchestra of over seventy musicians, and can also be acoustically adapted to accommodate Chinese opera by maneuvering a suspended acoustic canopy.

The playhouse, with seats in Mecanoo blue, is designed to host a variety of drama and dance performances. Flexibility is the core element in the design of this multifunctional space.

The recital hall has the most intimate atmosphere. With its asymmetrical composition and seating for around 400 across two levels, it is designed for chamber music. The seats have the same golden fabric as the concert hall and oak lines the walls. The upper part of this hall is enclosed by a circle of sound-absorbing curtains, allowing for the reverberation time within the space to be tuned to specific types of performance.

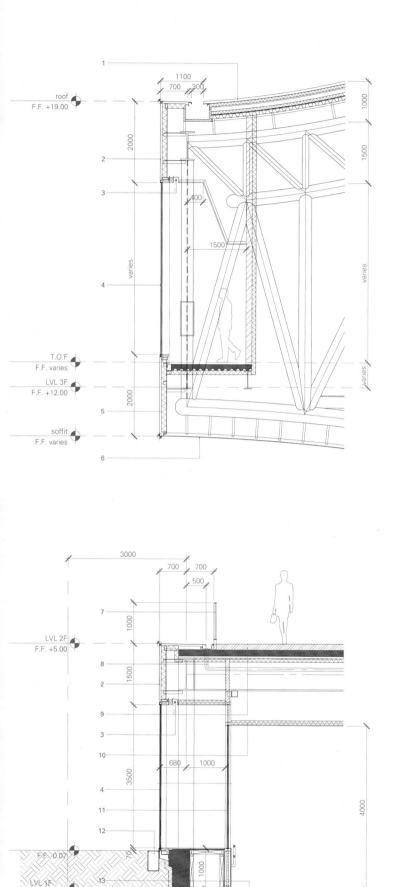

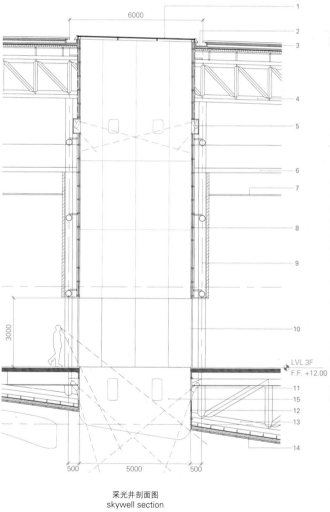

1. glass skylight
2. gutter
3. aluminum profiled sheet
4. roof truss
5. spotlight
6. maintenance catwalk
7. interior ceiling
8. aluminum plate finishing
9. interior finish
10. glass without frame
11. spotlight
12. metal panel
13. truss
14. steel skin with backing frame
15. view to banyan plaza

采光井剖面图
skywell section

详图1 detail 1

1. aluminum profiled sheet
2. aluminum plate finishing
3. reserve for lighting and blinds
4. I.G. with glass fin
5. aluminum plate finishing
6. steel skin with backing frame
7. steel balustrade
8. rainwater duct
9. PU floor
10. 3 gypsum board + mineral wool
11. double glass 13-20-9mm
12. line gutter
13. floating floor

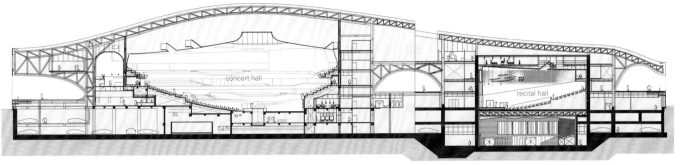

A-A' 剖面图 section A-A'

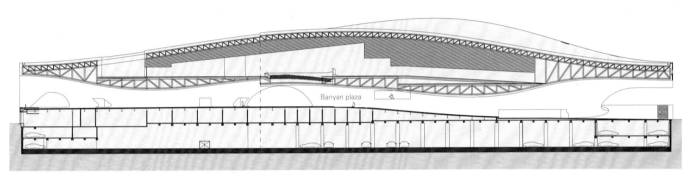

B-B' 剖面图 section B-B'

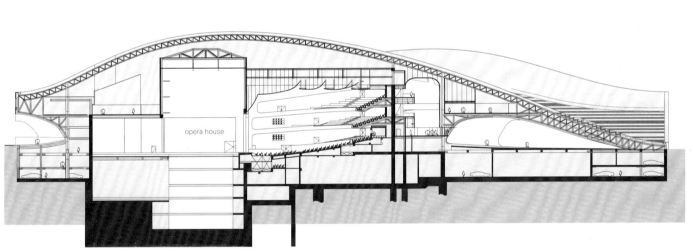

C-C' 剖面图 section C-C'

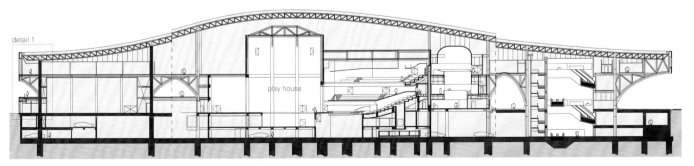

D-D' 剖面图 section D-D'

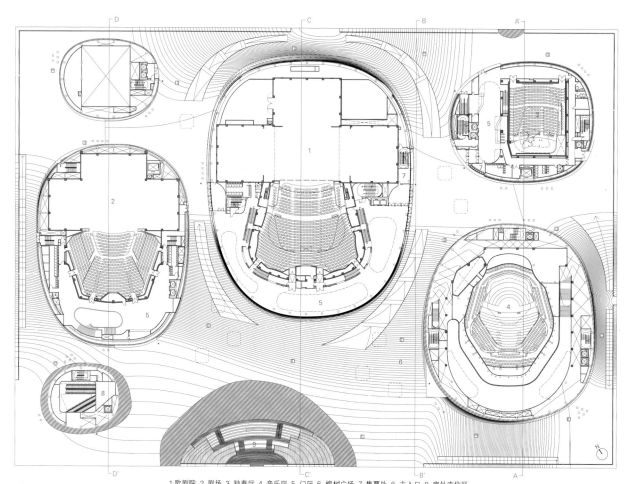

1.歌剧院 2.剧场 3.独奏厅 4.音乐厅 5.门厅 6.榕树广场 7.售票处 8.主入口 9.室外座位区
1. opera house 2. play house 3. recital hall 4. concert hall 5. foyer 6. Banyan plaza 7. ticket office 8. main entrance 9. outdoor seating area
三层 second floor

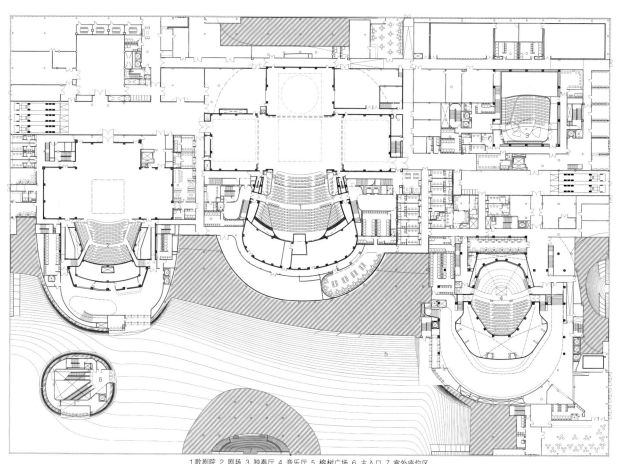

1.歌剧院 2.剧场 3.独奏厅 4.音乐厅 5.榕树广场 6.主入口 7.室外座位区
1. opera house 2. play house 3. recital hall 4. concert hall 5. Banyan plaza 6. main entrance 7. outdoor seating area
二层 first floor

项目名称：Kaohsiung Center for the Arts / 地点：No.1, Sanduo 1st Road, Fengshan District, Kaohsiung City, 83075, Taiwan / 事务所：Mecanoo architecten
当地合伙人、现场监理：Archasia Design Group / 合伙人负责人：Francine Houben / 项目建筑师：Nuno Fontarra / 项目指导：Friso van der Steen
设计团队：Aart Fransen, Bohui Li, Ching-Mou Hou, Danny Lai, Frederico Francisco, Jaytee van Veen, Joost Verlaan, Leon van der Velden, Magdalena Stanescu, Nicolo Riva, Rajiv Sewtahal, Reem Saouma, Sander Boer, Sijtze Boonstra, Wan-Jen Lin, Yuli Huang, William Yu, Yun-Ying Chiu
顾问：structural engineer – Supertech; mechanical engineer – Yuan Tai; electrical engineer – Heng Kai; acoustic consultant – Xu-Acoustique; theatre consultant – Theateradvies, Yi Tai; lighting consultant – CMA lighting; fire safety – Ju Jiang; organ consultant – Oliver Latry; roof and facade consultant – CDC; 3D consultant – Lead Dao; traffic consultant – SU International / 客户：Ministry of Culture
功能：theater complex in the Wei-Wu-Ying Metropolitan Park with a total capacity of 5861 seats; concert hall 1981 seats, opera house 2236 seats, playhouse 1210 seats, recital hall 434 seats, exhibition space of 800m², rehearsal + education halls for music and dance, 2 congress halls with 100 and 200 chairs and stage building workshops
用地面积：100,000m² (300,000m² building, 700,000m² park), metropolitan park 640,000m² / 建筑面积：36,000m² / 造价：366 million USD
设计时间：2007—2011 / 施工时间：substantial completion – 2010.4—2017.12; trial period, testing – 2018.1—9 / 开放时间：2018.10 /
摄影师：©Ethan Lee (courtesy of the architect) – p.136^{bottom}; ©Iwan Baan (courtesy of the architect) – p.133, p.134~135, p.136^{top}, p.137, 142^{bottom}, p.145^{top}; ©National Kaohsiung Center for the Arts (Weiwuying) (courtesy of the architect) – p.140~141, p.145$^{bottom-right}$; ©Shawn Liu Studio (courtesy of the architect) – p.128~129, p.139, p.142^{top} p.146~147

加建的不仅仅是空间

Addi
That Add Mo

若要对某一场所或建筑进行加建，除扩大原建筑的空间之外，还应添加更多的元素才算成功。仅仅扩大原建筑物的空间，既不妥当也无法满足使用需求。加建应该使建筑物本身更有意义。无论是原建筑还是加建的建筑，都应进一步彰显坐落其间的公司、组织或机构的整体风格。原建筑只能用有限的样式和建造技术体现风格，而新建筑则不拘泥于固定的样式，建造过程也更为先进。于是，在彰显坐落于建筑物中的空间之魅力方面，加建的建筑比原有建筑更胜一

An addition to a site, or to an existing building, can only be successful if more than physical space is added to the project. It is not enough to add more space or an extension of the existing, nor is it suitable to expand the existing. The new should add significance to the old, and both of them should further the overall narrative of the company, organization, or institution for which they are home. Whereas the architecture of the old had to express that narrative within limited styles and techniques of construction, the new architecture benefits from the liberal use of styles, and advanced construction processes. As a result, this allows

多伦多大学的丹尼尔斯大楼_University of Toronto Daniels Building / NADAAA
自然生物多样性中心_Naturalis Biodiversity Center / Neutelings Riedijk Architects
AGORA癌症研究中心_AGORA Cancer Research Center / Behnisch Architekten
CSN总部办公楼扩建项目_Expansion of CSN Headquarters / BGLA + NEUF Architectes
日内瓦鲁道夫·斯坦纳学校_Rudolf Steiner School of Geneva / LOCALARCHITECTURE
加建的不仅仅是空间_Additions That Add More Than Space / Phil Roberts

筹。本文讨论了五种可以促进建筑风格彰显的加建风格，并以新的视角审视不同职业领域建筑风格。其中一种加建方式体现了不同建筑组成部分的专业知识，一种展示了一部建筑史，还有一种则为希望解决问题的机构提供了理想的设计。每一种加建都是在原有风格中增添一抹新的色彩，而仅仅扩大空间则无法达到这样的效果。

additions to further the narratives of the entities within them in more compelling ways than the existing ever could on its own. This article looks at five additions with architecture that advances the overall narrative of the site, and expresses new ways of looking at their vocations. Where one addition expresses the expertise of its members, another displays a lesson in architectural history, while another gives an aspirational design to an institution dealing with an issue where hope is desired. Each addition adds a new chapter to the narrative, which could not have been achieved if only space was added.

加建的不仅仅是空间
Additions That Add More Than Space

Phil Roberts

20世纪之前，与某一职业相关的建筑物类型有限，样式刻板，建造流程毫无新意可言。这些局限性阻碍了建筑物加建、空间扩展和建筑构件的添加。如今，建筑物不必拘泥于过去的类型。教堂不需要再按照传统教堂的风格来建。现在，建筑物不再受限于所属职业领域的风格，多样的设计风格广受青睐。建造工艺的进步为建筑师提供了更多的创造方法。加建建筑不再受制于原有建筑的建造规则，而是以一种灵活的方式来讲述建筑的故事，进一步彰显场地的整体风格。

瑞士洛桑的AGORA癌症研究中心（192页），其加建建筑物呈现出不规则形状，旨在使建筑物周围的光照量达到最佳。Behnisch建筑事务所设计了一个由两部分组成的外立面，这两部分分别是一面坚固的外墙和一种带穿孔遮阳屏的轻型格架。

原有建筑物和加建建筑之间的空间，我们简单称之为AGORA，是一个中庭，作为在这两座建筑中工作的400名科学家和研究人员的休息室。加建建筑物中有着可俯瞰整个空间的内窗，与原建筑的内窗相协调。原有的和加建的建筑在一层连为一个整体，但是在上层空间则各自彰显特色。在AGORA中，唯一讲求功能性的空间是研究区域，因为这些空间具有实际功能。而加建结构中的所有其他空间都不具有功能性。

由于癌症研究是一项严肃的职业，因此其设计应严格遵循临床类型风格。原有的建筑及其对面街道的癌症病人护理大楼的外观有着明显的机构特色。相比之下，加建部分的设计则更为理想。原有建筑表达了在过去和现在我们对待癌症的态度；加建的部分也展现了我们现在应对癌症的方式，以及对未来研究的期望。秉承最大限度利用阳光的动态设计理念，加建结构为原本代表悲观沮丧的职业代入一种乐观的态度。

就像AGORA建筑告诉我们要对癌症研究进展充满希望一样，多伦多大学的约翰·H.丹尼尔斯建筑、景观和设计学院的建筑也令人振奋（154页）。该案例中，Spadina Cresent场地的既存建筑老诺克斯学院，没有容纳设计功能，而是多年不对外开放。该工程所要改造的老建筑离老学院不足1km远，由于该学院搬迁到Spadina Cresent，给了老学院所在的地方以希望，这里长期以来一直是一个环岛，废弃的大学建筑就位于环岛的中心位置。由NADAAA建筑事务所承建的加建工程现在将诺克斯学院包含在内，人们可以360°欣赏到

Prior to the early 20th century, the vocation of a building was stuck within limited typologies, strict styles, and the construction processes of the day. Those limitations governed expansions, extensions and additions. Today, buildings do not have to adhere to the typologies of the past. A church no longer has to look like a church. With no overriding style relative to vocation to dictate what buildings look like today, more diversity of design is appreciated. The advancements in construction technology has given architects more ways to create. An addition is no longer beholden to the rules that governed the existing, but is free to tell its story in a way that furthers the narrative of the site as an ensemble.

At the AGORA Cancer Research Center in Lausanne, Switzerland (p.192), the addition's irregular form, and the favourable orientation optimize the amount of sunlight around the perimeter of the building. Behnisch Architekten designed a facade with two parts: a solid exterior wall, and a lightweight lattice of perforated shading screens.

The space between the existing and addition, simply called the AGORA, is an atrium which functions as a lobby for the 400 scientists and researchers who use both buildings. The interior windows of the addition overlooking the space are proportional to those on the existing. Both buildings are linked on the ground floor as one, but make their differences more pronounced above ground. In the AGORA, the only space that follows function is the research areas, because of their pragmatic requirements. All other spaces in the addition are formed in ways that deviate from function.

With cancer research being such a serious vocation, it is understandable that it would exist in a building with a heavily clinical typology. Whereas the existing building, and the cancer care buildings across the street have an obvious institutional look, the addition is more aspirational. The existing expresses how we dealt with cancer before, and how we deal with it now. The addition also expresses how cancer is dealt with now, and the hope that exists for the future. With a dynamic design that is oriented to maximize sunlight, the addition provides an optimistic perception for what a typically a gloomy vocation.

The same way the architecture of AGORA teaches us to be hopeful for the advancement in cancer research to come, the University of Toronto's John H Daniels Faculty of Architecture, Landscape, and Design building inspires. (p.154) In this case, the existing building on the Spadina Cresent site, the old Knox College, did not house the design program, but remained inaccessible for years. The program's existing building was less than a kilometer away, but its move to Spadina Cresent has given life to what was for too long, just a roundabout with a defunct university building at its core.

它的全貌。每个视角都可以给观赏者提供一堂建筑史课。大家在有轨电车上（行驶路线：Spadina Cresent的西半部至东段）观赏，可以领略到丹尼尔斯大楼的最佳景观。该学校教授的三门学科的案例在建筑场地中都能找到，这使得它成为多伦多过去和现在建筑环境的样本。

　　从北面看，加建建筑的整个立面都是落地窗，这些窗户对称设置，形成当代风格的表面。研究生工作室、本科生公用办公桌空间和数字制作实验室的活动都一览无余。这是一座以透明公开的方式与周围世界交流的建筑，互联网和社交媒体使公众比以往任何时候都更容易了解该职业的意义。建筑两侧的护道形成了噪声缓冲器，以防止来自有轨电车轨道的噪声和车辆轰鸣声。当学生在数字制作实验室工作时，护道会提供一个缓压区域，使他们可以从室内走出来，进入室外庭院来缓解压力。

　　南立面是翻新的诺克斯学院建筑，它代表着一个世纪前的建筑风格，其层次结构、图案和风格都是强行规定的，完全复制以前的建筑风格。尽管加建建筑与诺克斯学院迂腐老套的文体建筑课毫无关联，但新旧建筑的共同点是都是景观建筑，并尊重设计比例。

　　东西两侧的立面是新旧建筑连接的地方，代表了我们今天与建筑互动的方式。在东侧，一个广场欢迎着多伦多大学校园的学生们，学生们可以通过一条狭窄的小道通往西侧住宿区。新旧建筑连接之处即是丹尼尔斯大楼的主入口。此外，主入口设在加建建筑中而不是原有建筑中不仅具有建筑意义，而且具有象征意义。对于今天学习建筑和设计的学生来说，过去仅仅是现在的参考，但是他们进入专业领域是为了解决今天的专业问题，而非过去。这也就是为什么让他们从加建建筑这一侧进入是合适的。

　　加建建筑的屋顶能让阳光深深地照进建筑物的内部，而其他几座建筑则鲜能做到这一点。加建建筑内部的每个表面、空间、纹理、形状和形式，以及建筑的外观，都致力于启发下一代设计师，为其提供设计灵感。这座建筑本身就是这样一个活生生的例子，学生就是在这样的建筑中接受教育、学习和实习。这是一座鼓励创造力的建筑，并尽力在内部设计上与普通学术空间彻底脱离。这些是为人们去建筑学校学习所创建的空间，不是为节省建筑成本而潦草完工的复制品。加建建筑和诺克斯学院一并成为伟大建筑学经久不衰的教程。

The addition by NADAAA, which now includes Knox College, can be observed from 360 degrees. Each angle gives the viewer an architectural history lesson. The best view of the Daniels Building is on the streetcar, which glides around the western and eastern halves of Spadina Cresent. Examples of all three disciplines taught at the school can be found on the site, which makes it a sample size of Toronto's built environment, past and present.

Coming from the north, the addition has a symmetrical contemporary face of floor to ceiling windows across the whole elevation. The activity of the grad studios, undergrad hot desk spaces, and the digital fabrication lab are in full view. This is representative of an architecture communicating with the world around it in a transparent way, with the internet and social media making it easier than ever for the public to see what the profession is about. The berms on either side of the elevation form a sound buffer from the traffic and clanking coming from the streetcar rails. When students are working in the digital fabrication lab, the berms create an area where they can come outside in the construction court and relieve stress.

The southern facade, which is the restored Knox College building, is representative of architecture as it was a century ago, dictating hierarchies, patterns, styles, and obsessed with replicating what came before. Whereas the addition is indifferent to the pedantic stylistic lessons of Knox College, what both have in common is landscape architecture and the respect for proportions.

The eastern and western facades are where both buildings meet, and represent how we interact with architecture today. From the east, a plaza greets students from the UofT campus, while a smaller path provides a connection to the residential area to the west. Where the addition and the existing meet is where the actual main entrance to the Daniels Building can be found. The location of the main entrance on the addition and not the existing not only makes sense in terms of architecture, but in terms of symbolism as well. For the architecture and design student of today, the past is just an ever-present reference, but they enter the profession looking to solve the problems of today, not of yesterday. That is why it is fitting that they enter through the addition.

The roof of the addition is positioned to permit the passage of sunlight deep into the building at angles that few can experience in other buildings. Each surface, space, texture, shape, and form found on the interior of the addition, as well as the architecture on the exterior, serves to inspired the next generation of designers. They are taught, study, and build

并不是所有的加建建筑都像AGORA和丹尼尔斯大楼一样与原建筑形成鲜明的对比。瑞士Confignon鲁道夫·斯坦纳学校的加建建筑（218页）显得更为微妙，加建建筑与20世纪80年代中期由让-雅克·屈米设计的原有建筑结构相同，不同点在于外立面的颜色。该建筑在原有七个构造独特的教室基础之上，增加了一个新的楼层，可以俯瞰内部庭院。

木材是所有加建教室所用的主要材料，被作为结构构件和表层饰面。宽阔的窗户和天窗让阳光充足地洒进每个教室，增强了年轻人的学习体验。与原有混凝土结构中较单调的照明相比，这是一个令人耳目一新的变化。由于教室形状独特，加建的教室里很少有光线暗的角落，走廊上方倾斜的屋顶是一个小的设计特色，使增加的新楼层保持明亮。卡萨工作室的木匠师傅参与了这个项目，通过使用三维制造技术与建筑师紧密合作，加建出与楼下原有的几个楼层相同的形式。

鲁道夫·斯坦纳学校加建建筑的屋顶沿用太阳能电池板，是来这所学校学习的年轻学生在可持续发展设计方面一个可借鉴的案例。学校原来的混凝土结构易产生温室气体，与之不同，加建的部分采用了暖色调、典雅庄重的更能持续使用的木材结构。在加建建筑施工期间，参与设计的学生定期到工地实时考察，以一种在AGORA和丹尼尔斯大楼中无法实现的方式，来比较新旧建筑的不同。在鲁道夫·斯坦纳学校，新旧建筑呈现类似的形式，建筑设计采用不同的材料，所以参与其中的学生能更好地理解可持续建筑的重要性。

掌握使用某种材料的专业知识，是加拿大最大建筑工会之一的总部扩建项目（206页）向在那工作的人以及广大公众所展示的主要内容。两家魁北克建筑公司——BGLA和NEUF建筑事务所联合，设计了该总部的加建建筑，两所建筑事务所想在CSN总部的设计中明确区分原有建筑和加建建筑。人们普遍认为野兽派建筑太冷，令人难以接近，加建的建筑彰显了CSN的价值观，即自治、自由、团结。

中庭通过泥土色的表面和大量木材的使用，展现了魁北克建筑业的专业知识。BGLA倾向于在大多数项目中使用木材作为其突出特征，因为木材是魁北克的文化遗产。室外有一个与中庭相连的广场，是供工作人员聚集的共用空间。所有室内办公室均设有可俯瞰

in a building that is itself an example of what is possible. A building that encourages creativity, and internally strives to be a radical departure from generic academic spaces. These are the spaces that people go to architecture school to create, that are not construction cost induced replicas. The addition, along with Knox College, is a constant tutorial of great architecture.

Not all additions express a stark contrast with the existing like the AGORA and Daniels Building. The addition for the Rudolf Steiner School in Confignon, Switzerland (p.218) is more subtle. The addition carries the same organic form as the existing that was designed by Jean-Jacques Tschumi in the mid 1980s, but the difference is in the color of the facade. The LOCALARCHITECTURE addition is a new storey of seven uniquely formed classrooms overlooking an inner courtyard. Wood is the predominant material in all the added classrooms, and is used as the structure and surface finish. The wide windows and skylights allow sunlight to enter each classroom in abundance, enhancing the learning experiences of the youth. It is a refreshing change from the drab lighting in the existing concrete structure. There are few dark corners in the added classrooms because of their form, and the slanted roof above the corridor is a small design feature that keeps the new level bright. Ateliers Casai, the carpenters involved in the project, worked closely with the architects by using three-dimensional fabrication technology to give the addition the same form as the existing levels below.

The addition for the Rudolf Steiner School, with its solar panels along the roof, is an example of sustainable design for the young students attending the facility. Compared to the existing part of the school, with its greenhouse gas producing concrete structure, the addition expresses the warmth and modesty of the more sustainable wood structure. The students, who were given regular tours of the addition during construction, get to compare the old and the new, in a way that is not possible at the AGORA or the Daniels Building. At the Rudolf, both sections have similar forms, but different materials, so the students can see and understand the importance of sustainable construction.

The expertise that comes from mastering the use of certain materials is what the addition to the headquarters of one of Canada's largest construction trade union (p.206) expresses to those who work there, and to the public at large. Designed by a consortium of two Quebec architecture firms, BGLA and NEUF Architectes wanted to make a clear distinction between the addition and the existing at the headquarters of the Confédération des syndicats nationaux (CSN). The brutalist architecture of the existing building is seen as too cold and unapproachable. The addition expresses the values of the CSN, which are autonomy, freedom, solidarity.

The atrium, with its earth tone surfaces and heavy use of wood, showcases the expertise of Quebec's construction sector. BGLA favours using wood as a prominent feature in most of their projects, since it is a material that carries a cultural

中庭的窗户，并且三条对角通道将扩建部分与原有通道连接起来。

不团结，工会就毫无用处。加建建筑旨在表达贸易成员之间的象征性纽带。并不是说原有的建筑物未能在工会会员之间树立团结理念，因为对加建建筑的强烈需要就证明了原建筑实现了这一理念。不同之处在于，加建建筑更加清楚地展现了这种团结理念。与多伦多面向建筑系学生的丹尼尔斯大楼相似，扩建的CSN总部定期提醒其成员可以做些什么。

最后，由诺伊林斯·里迪克建筑师事务所设计扩建的荷兰莱顿自然生物多样性中心（174页）向人们展示了该建筑是如何延续自然历史的叙述的。作为4200万件文物的收藏场所，自然生物多样性中心的建筑代表了荷兰形形色色的历史。加建结构容纳了新的博物馆，以展示更多的文物，这对小学生实地参观建筑和广大公众都是有益的。加建建筑还供200多位研究生物多样性、水资源、粮食供应和气候变化的研究人员使用。

自然生物多样性中心融多种建筑风格于一体，一些是原有建筑，另外一些是加建建筑。每座建筑物的形式、形状、建材和外观都有明显不同。原有建筑物的空间结构使其无法大量展示手工艺品，但通过加建，自然生物多样性中心可以使更多人参观其研究过程。当研究人员检查生物标本时，中庭的参观者可以在中央楼梯的顶部或走廊中观看。带有白色立方体透镜图案的中庭是每座建筑物独特视觉语言的汇聚点。中庭的天花板上有着相同的几何图案。所有的展览馆都是用粗糙凿成的石头砌成的，这些石头从外至内包裹着中庭，在材料的使用上具有连续性。在这一地质主题建筑中，艾里斯·范·荷本创作的263条灵感来自自然的花卉饰条装饰在石头纹路之间。在博物馆内，贴有100幅照片和图片的墙板讲述了地球的自然历史。

在这五个项目的每个项目中，原有建筑和加建项目都是连在一起的，彼此之间都没有利用开放空间来真正将自身与另外一方区分开来，但是无论如何它们还是存在差别的。每个项目都不仅仅是扩大、扩展和增加空间，同时还为用户和公众提供了更多东西，即更多的希望、灵感、理解、团结和教育。最重要的是，加建为经历了这些加建过程的人们的生活创造了更大价值。

heritage in Quebec. There is an agora on the exterior which links to the atrium, where the common areas for the workers are situated. All interior offices have windows overlooking the atrium, and three diagonal passageways connect the addition with the existing.

Without solidarity, a trade union is of no use. The addition is designed to express the symbolic bonds between trade members. Not that the existing building failed to build unity among the union's membership, because the very need for an addition is proof of their unity. The difference is that the addition is a clearer representation of that togetherness. Similar to the Daniels Building for architecture students in Toronto, the CSN's addition is a regular reminder of what its members can do.

Finally, there is Neutelings Riedijk Architects addition to Naturalis Biodiversity Center in Leiden, Netherlands (p.174), and how it continues the narrative of natural history. As the home to 42 million artefacts, the architecture of Naturalis represents the diverse history of the Netherlands. The addition houses a new museum to put more artefacts on display, which is a benefit to school children on field trips and the general public. Within the addition is also space for over two hundred researchers who study biodiversity, water, food supply, and climate change.

Naturalis is many buildings in one ensemble, some existing and others additions. Each building has a noticeably different form, shape, material and appearance. The existing buildings did not allow for the extensive exhibiting of artefacts, but with the additions, Naturalis can expose more people to its research. Visitors in the atrium can watch as researchers examine the corpses of creatures, and at the top of the centrally located stairs, waits a T-Rex in a gallery.

The atrium, with its pattern of lenses over a white cubic mass, is the gathering point where the separate visual languages of each building meet. The same geometric pattern is found on the ceiling of the atrium. All of the exhibition galleries are contained within blocks of rough-hewn stone, which goes from the exterior and wraps around inside the atrium, providing material continuity. Within this geological theme are 263 nature-inspired floral friezes by Iris van Herpen between the courses of stone. Inside the museum, 100 wall panels of photography and images tell the natural history of the globe.

In each of these five projects, the existing and the addition are physically connected. Neither has the benefit of an open space between them to truly distinguish themselves from the other, yet they do anyways. More than an expansion, extension, and an addition of space, each project gives its users and the public something more. More hope, inspiration, understanding, solidarity, and education. Above all, value is added to the lives of those who experience these additions.

多伦多大学的丹尼尔斯大楼
University of Toronto Daniels Building

NADAAA

NADAAA 为富有悠久历史的建筑增添了带有错落有致屋顶、犹如雕塑般的空间
NADAAA added a sculptural extension with a jagged roof to a historic building

位于多伦多大学的丹尼尔斯大楼体现了城市设计和可持续发展思想相结合的设计理念。作为约翰·H.丹尼尔斯建筑、景观和设计学院新的办公学习场所，丹尼尔斯大楼旨在为更多的学生和社团提供关于建筑环境讨论的氛围。

此项目地处多伦多为数不多的几个圆形地块之一的中心，坐落于多伦多大学的西南角，在多年封闭之后终于向公众开放。此项目重塑了这栋古老且几乎被人遗忘的建筑的恢宏，同时增加了一个新的结构。与原建筑融为一体的新的加建部分犹如雕塑，由金属、混凝土和玻璃建造，其东侧和西侧表面为折叠的混凝土板，而朝北的立面则是黑框幕墙。

南北向的轴线体现了与城市之间的象征性关系，行人通行路线则形成了东西向的轴线。在西侧的边缘处，一条精心设计的拱廊强调了附近社区住宅的规模。同时，东侧的公共广场与校园有着紧密的联系。改建后的建筑为更多的活动提供了场地，包括为行人和骑行者设计的交通流线。值得一提的是，在其众多可持续设计特色中还包括一个场地雨水管理系统，它同时也使这座遗产建筑恢复了生机与活力。

此建筑的设计给人们展示了这样一个案例，在这个案例中，教育方面的问题仍需要在一个现实环境中进行研讨。专家、评论家、教师、从业人员和学生都在这个环境中生活，同时也检验着这个环境，是鉴赏这座建筑的主角。这可能是为数不多的公众能够和建筑及其设计者

在更高层次上进行沟通交流的机会，从而给建筑师带来更多的挑战和责任，建筑师需要更加深入地进行建筑细节的探讨。正因如此，建筑被设计成一个教学工具，将可持续设计元素和学校项目融合其中，向学生和公众展示。

The Daniels Building at the University of Toronto embodies a holistic approach to urban design and sustainability. As the new home for the John H Daniels Faculty of architecture, landscape, and design, its purpose is to engage students and the broader community in dialogue about the built environment. At the center of one of Toronto's few circular parcels, the project anchors the southwest corner of the university and opens the circle to the public after years of inaccessibility. It restores the historic and forgotten building to its original grandeur while also integrating a new addition. Adjoining the original structure is a sculptural mass made of metal, concrete and glass. The east and west sides are wrapped in folds of concrete, while the north-facing elevation features a black-framed curtain wall.

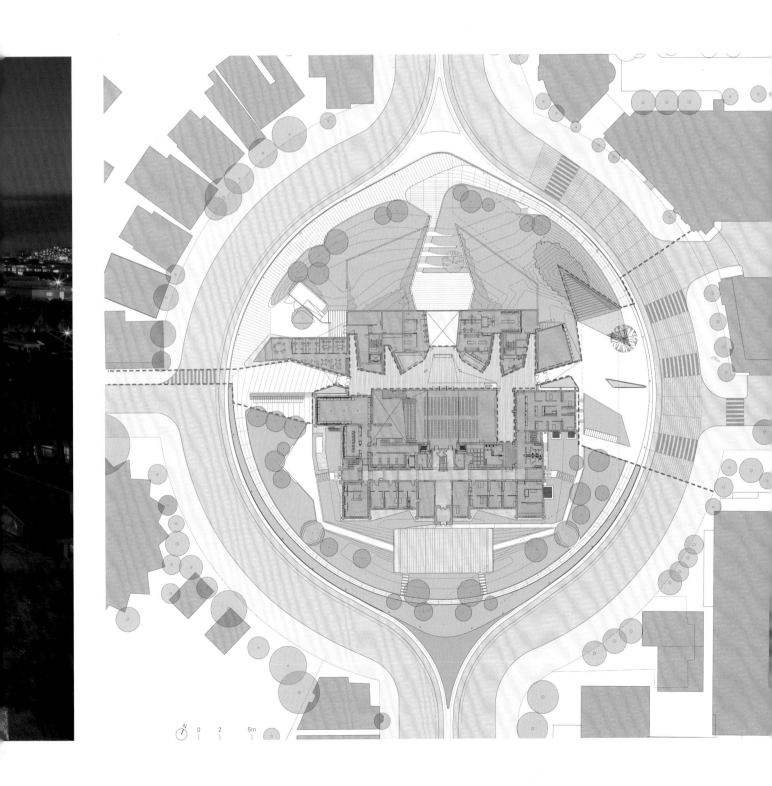

1950年3月 March 1950

集会大厅 Convocation Hall

左图：1956年4月，索尔克脊髓灰质炎疫苗包装
右图：1945年9月，青霉素生产中的运输模型文化
left: April 1956, packaging of Salk polio vaccine
right: September 1945, transporting mold cultures in the production of penicillin
©Sanofi Pasteur Canada (Connaught Campus) Archives

南立面 south elevation

东立面 east elevation

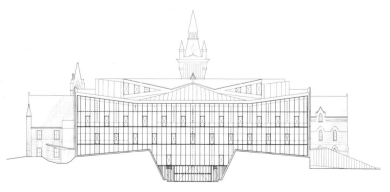

北立面 north elevation

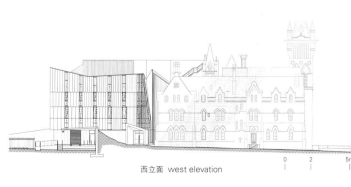

西立面 west elevation

详图——西墙
detail 1_west wall

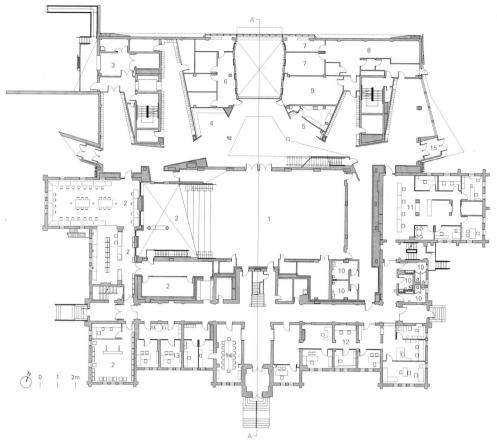

一层 ground floor

1. 主厅
2. 图书馆
3. 卸货/储存区
4. 休息室
5. 咖啡厅/厨房
6. 3D打印区
7. IT办公室
8. 测绘室
9. 激光切割室
10. 卫生间
11. 学生服务区
12. 行政套房
13. 办公室
14. 董事办公室
15. 主入口

1. principal hall
2. library
3. loading/storage
4. lounge
5. cafe/kitchen
6. 3D printing
7. IT office
8. plotting room
9. laser cutters
10. restroom
11. student services
12. admin suite
13. office
14. boardroom
15. main entrance

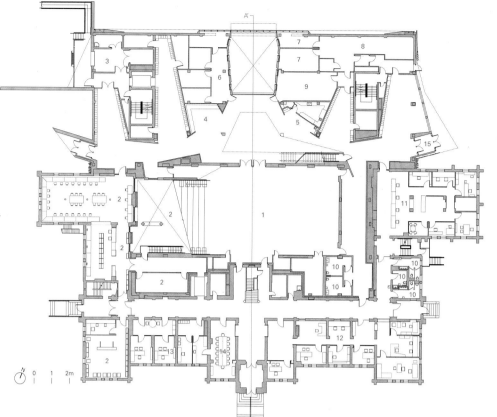

地下一层 first floor below ground

1. 数字制作室
2. 金属车间
3. 焊接室
4. 模型/组装室
5. 木材/普通材料车间
6. 喷漆室
7. 真空成型机
8. 员工室/水槽柜
9. 车间办公室
10. 小组学习室
11. 普通收藏区
12. 画廊
13. 机械室
14. 卫生间
15. 压缩机/集尘室
*未标注区域为机械室

1. digital fabrication
2. metal shop
3. welding
4. mock-up/assembly
5. wood/general shop
6. spray booth
7. vacuum former
8. staff/sink lockers
9. workshop office
10. group study
11. general collections
12. gallery
13. mechanical room
14. restroom
15. compressor/dust collection
*unlabeled areas are mechanical

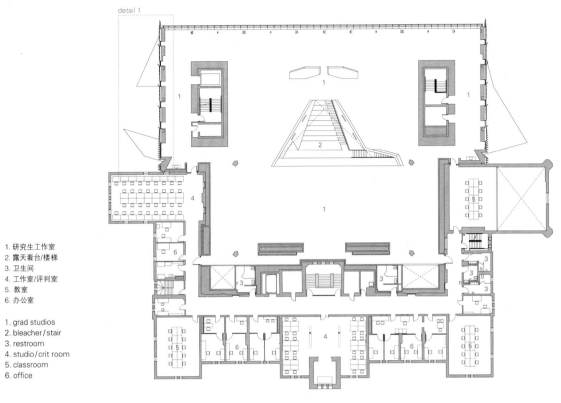

1. 研究生工作室
2. 露天看台/楼梯
3. 卫生间
4. 工作室/评判室
5. 教室
6. 办公室

1. grad studios
2. bleacher/stair
3. restroom
4. studio/crit room
5. classroom
6. office

三层 second floor

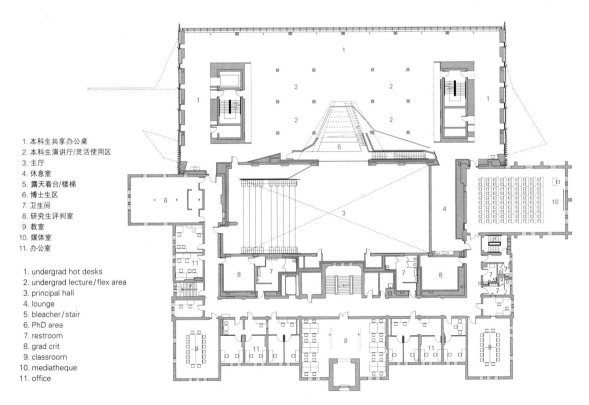

1. 本科生共享办公桌
2. 本科生演讲厅/灵活使用区
3. 主厅
4. 休息室
5. 露天看台/楼梯
6. 博士生区
7. 卫生间
8. 研究生评判室
9. 教室
10. 媒体室
11. 办公室

1. undergrad hot desks
2. undergrad lecture/flex area
3. principal hall
4. lounge
5. bleacher/stair
6. PhD area
7. restroom
8. grad crit
9. classroom
10. mediatheque
11. office

二层 first floor

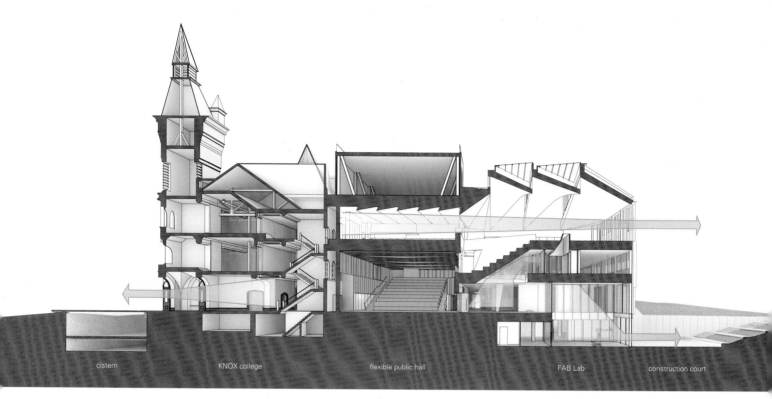

A-A' 剖面图 section A-A'

1. 玻璃幕墙 2. 遮阳帘 3. 制冷"白"屋顶 4. 自动开启窗 5. 自动烟囱通风 6. 北向天窗 7. 置换通风 8. 低采光电能密度设计
9. 任务采光 10. 太阳能屋顶 11. 空心预制混凝土板 12. 高反射室内空间 13. 中央场地水箱

1. glazed curtainwall 2. sunshades 3. cool "White" roofs 4. automated operable windows 5. automated stack ventilation 6. north-facing clerestory windows 7. displacement ventilation 8. low lighting power density design 9. task lighting 10. solar-ready roof 11. voided precast concrete slab 12. high-reflectance interiors 13. central site cistern

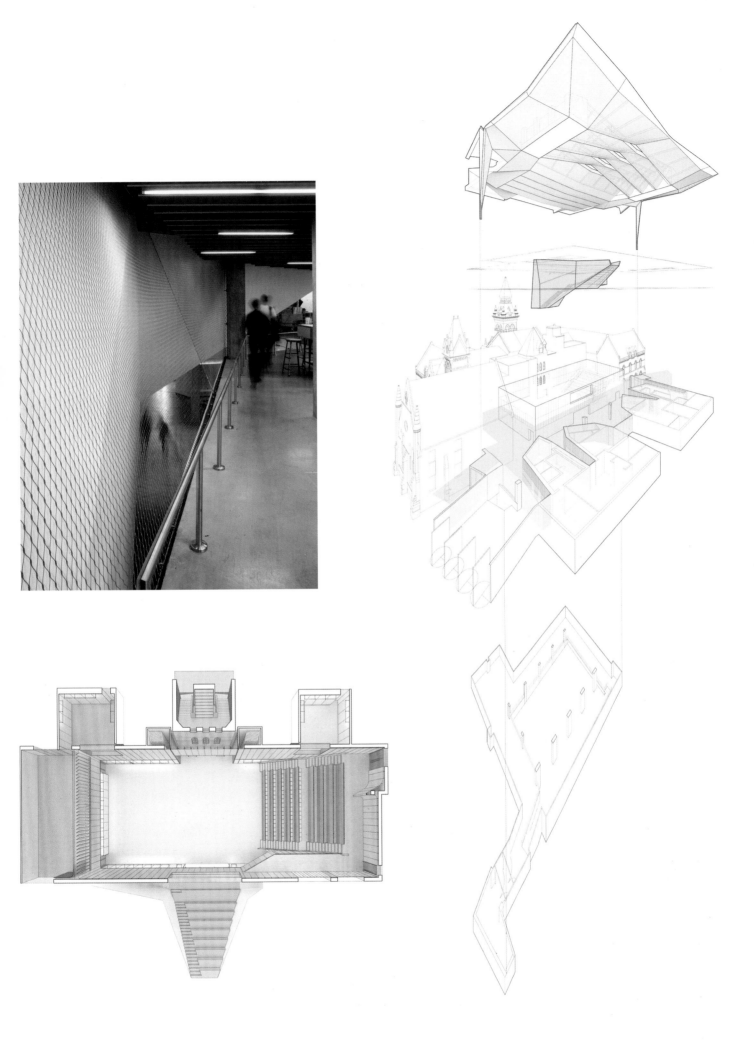

The north-south axis characterizes symbolic relationships to the city, while the east-west axis is activated by pedestrian traffic. On the western edge, a discreet arcade addresses the residential scale of the adjacent neighborhood. Meanwhile, a public plaza to the east creates a prominent relationship with the campus. The renewed site invites activity, with circulation for pedestrians and cyclists. Extensive sustainable features including a noteworthy approach to site stormwater management while simultaneously bringing a heritage building back to life.

The design of this building presented a case where problems of pedagogy come face to face with a physical environment

that is inhabited and tested daily by an audience of experts, critics, teachers, practitioners, and students, the very protagonists of the medium. It is perhaps one of the few occasions where the audience is engaging with the building and its authors at a higher level, making it an added challenge – and responsibility – to speak to architectural questions with a greater degree of nuance. As such the building was designed to become a pedagogical tool, integrated into the curriculum with both sustainable design elements and school programs on display both to students and the public.

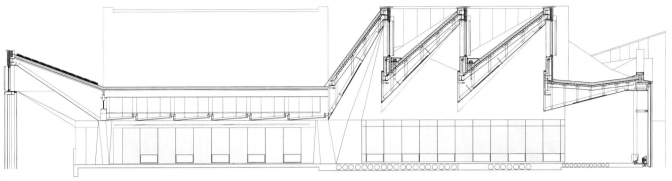

屋顶剖面详图 roof section detail

项目名称：University of Toronto Daniels Building / 地点：One Spadina Crescent, Toronto, ON, Canada / 事务所：NADAAA / 负责人：Katherine Faulkner, Nader Tehrani / 项目经理：Richard Lee, Tom Beresford / 合作建筑师：John Houser, Amin Tadj, Tim Wong, Alda Black, Marta Guerra, James Juricevich, Parke Macdowell, Nicole Sakr, Dane Asmussen, Laura Williams, Peter Sprowls, Noora Al Musallam, Tammy Teng, Wesley Hiatt, John Mars, Mazyar Kahali, Matthew Waxman, Luisel Zayas / 合作建筑师负责人：Adamson Associates Architects Principal: Claudina Sula

项目团队：Jack Cusimano, Tina Leong, John McMillan, Martin Dolan, Zbigniew Jurkiewicz, Michael Lukachko, Zale Spodek, Gilles Leger, George Georges, Ke Leng Tran / 客户：Daniels Faculty of Architecture, Landscape, and Design, University of Toronto / 遗产顾问：ERA Architects / 结构：Entuitive Corporation / 建筑围护结构顾问：Entuitive Corporation / 电气、数字、AV、照明设计：Mulvey Banani International, Inc. / 机械、管道：The Mitchell Partnership / 声学设计：Aercoustics Engineering Ltd / 土木工程：A. M. Candaras Associates, Inc. / 景观：Public Work / 硬件：Upper Canada Specialty Hardware, Ltd. / 施工管理：Eastern Construction Company Ltd / 家居设计与建造：Daniels Faculty / 面积：14,400m² / 用地面积：11,441m² / 建筑面积：14,400m² / 总楼面积：14,400m² / 设计开始时间：2011 / 施工时间：2014—2018 / 摄影师：©Nic Lehoux (courtesy of the architect)

自然生物多样性中心
Naturalis Biodiversity Center

Neutelings Riedijk Architects

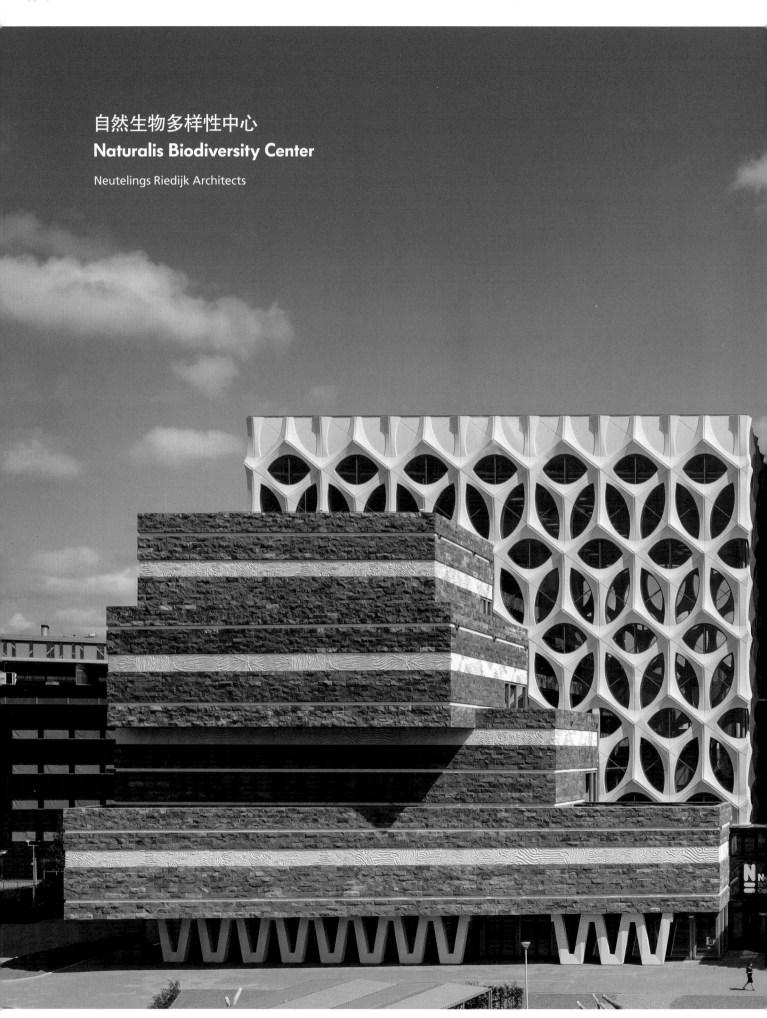

Neutelings Riedijk 建筑师事务所对荷兰一家著名的国家研究机构进行了面向未来的翻新
Neutelings Riedijk Architects' future-proof renovation of a well-established national research institute in the Netherlands

该中心是荷兰国家生物多样性研究机构,成立于1820年,由国王威廉一世创立于莱顿。该研究机构历史悠久,经验丰富,在过去的十年间飞速发展,因此迫切需要进行翻新。参观者的数量迅速增加到每年40万。随着参观者数目的增加,中心也扩大了规模,加建后可容纳4200万观众同时参观,这使得该中心跻身世界前五名。

该项目占地约38 000m²,其中18 000m²为翻新工程,20 000m²为新建工程。先进的新设施可容纳200多名研究人员。他们的研究成为人们关注的焦点,为解决全球问题提供了解决方案,其中包括气候变化、地球生物多样性的下降、食物供应和水质问题。中心设施和馆藏使研究人员能够提供最高水平的解决方案。同时,新博物馆为向公众展示自然丰富资源和美丽景观提供了契机。

该研究机构的新设计构成了原有建筑物和新建建筑物的可持续组合,每种活动均以特定形式进行。中央中庭将研究机构的各个部分连接在一起,包括原有的办公室和仓库以及新建的博物馆和实验室。中庭的设计包括一个互锁分子形式的三维混凝土结构,犹如椭圆形、三角形和六边形的花边。从圆形窗户进入的经过过滤的光线就像是"玻璃冠",增强了该空间的纪念意义,科学家、教职员工、学生和家庭将在这里相遇。

餐厅、商店和展厅等公共场所位于一层,路人可以看到生物学家检查冲上岸的鲸鱼。通往展览空间的主楼梯就像一条山路,顶部逐渐变窄,只留下足够的空间容纳特丽克斯——有着6600万年历史的霸王龙标本,它在恐龙时代画廊中享有很高的地位。

展厅的外部是水平结构的石块,刻意模仿了地质结构。多种石材的使用创造出美丽的火花。石层之间装饰着荷兰著名时装设计师艾里斯·范·荷本设计的白色混凝土饰条。受中心邀请,她设计了共263块面板,灵感来自馆内收藏品的自然形状。由于使用了为中心开发的特殊技术,这些面板看起来像丝绸一样光滑,很像范·荷本为凯特·布兰切特、碧昂斯和Lady Gaga等名人设计的创新礼服的布料。

在博物馆内,以舞台灯光设计、家具设计和带有精美花卉与动物图案的织物设计而闻名的荷兰设计师托德·邦杰,展示了近100种醒目的彩色壁板。这些是将摄影和绘画融为一体的视觉体验,揭示了自然界的奇观。

Naturalis is the national research institute for biodiversity in the Netherlands, dating from 1820, and founded by King Willem I in Leiden. The institute, which boasts a long and rich history, experienced exponential growth in the last decade which led to an urgent necessity to renovate. The number of visitors increased rapidly to 400,000 per year. The new future-proof Naturalis brings the growing collection of 42 million objects together, putting it among the top five in the world.

The project covers a total of around 38,000m² of which 18,000m² is renovation and 20,000m² is new construction. New state of the art facilities accommodate more than 200 researchers whose studies are placed at the center of the attention, contributing solutions to global issues including climate change, the decline of biodiversity on earth, food supply and water quality. The Naturalis facilities and the collection enable researchers to contribute solutions at the highest level. At the same time, the new museum offers the chance to show the public the wealth and beauty of nature.

The institute's new design forms a sustainable ensemble of existing buildings and new-build, with each activity housed in a specific form. The central atrium connects the various parts

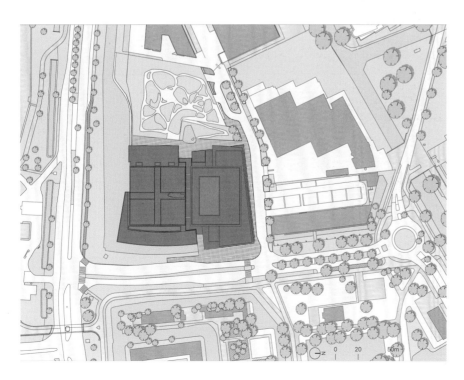

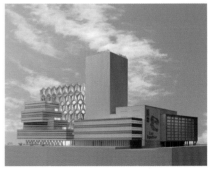

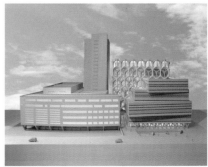

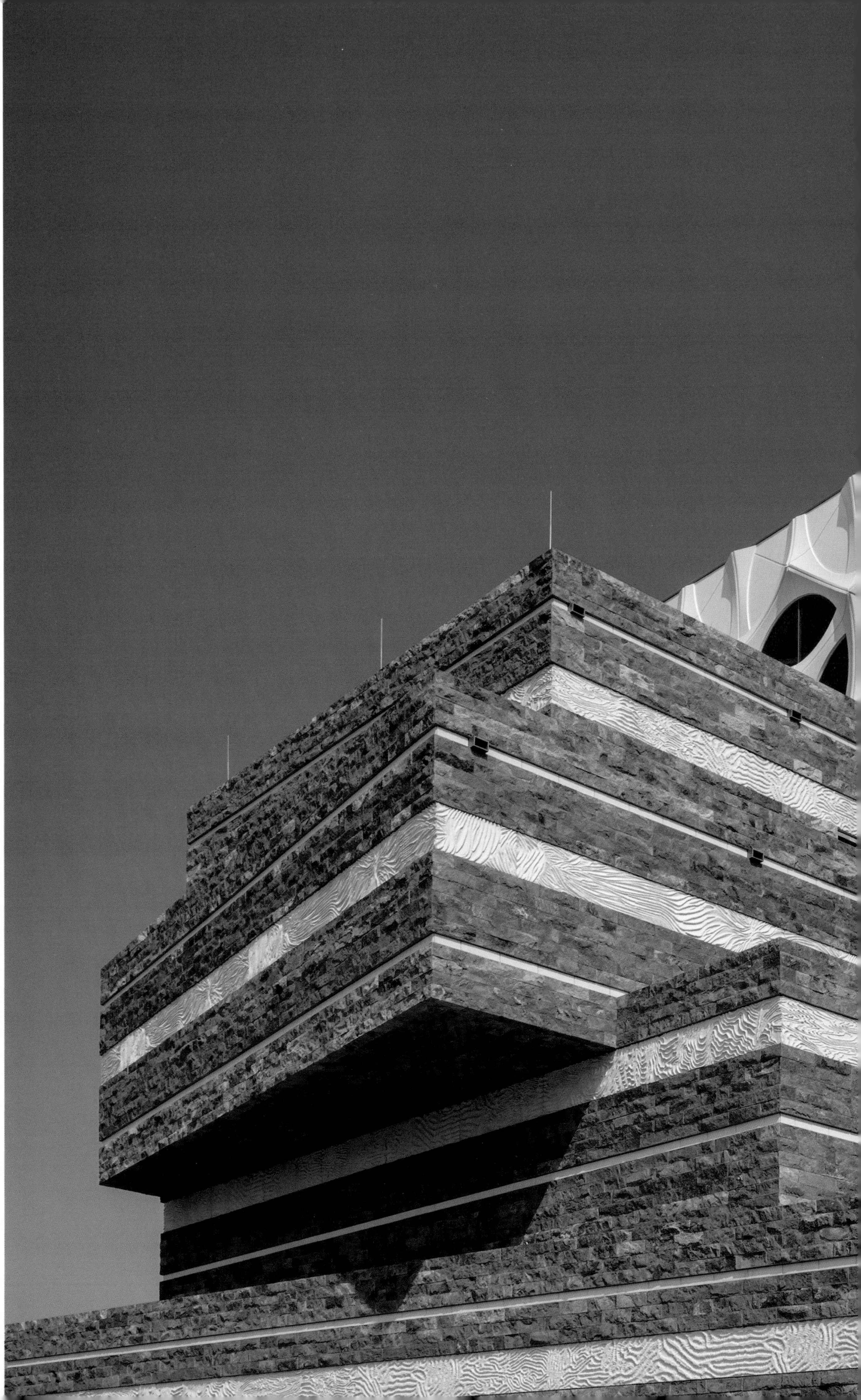

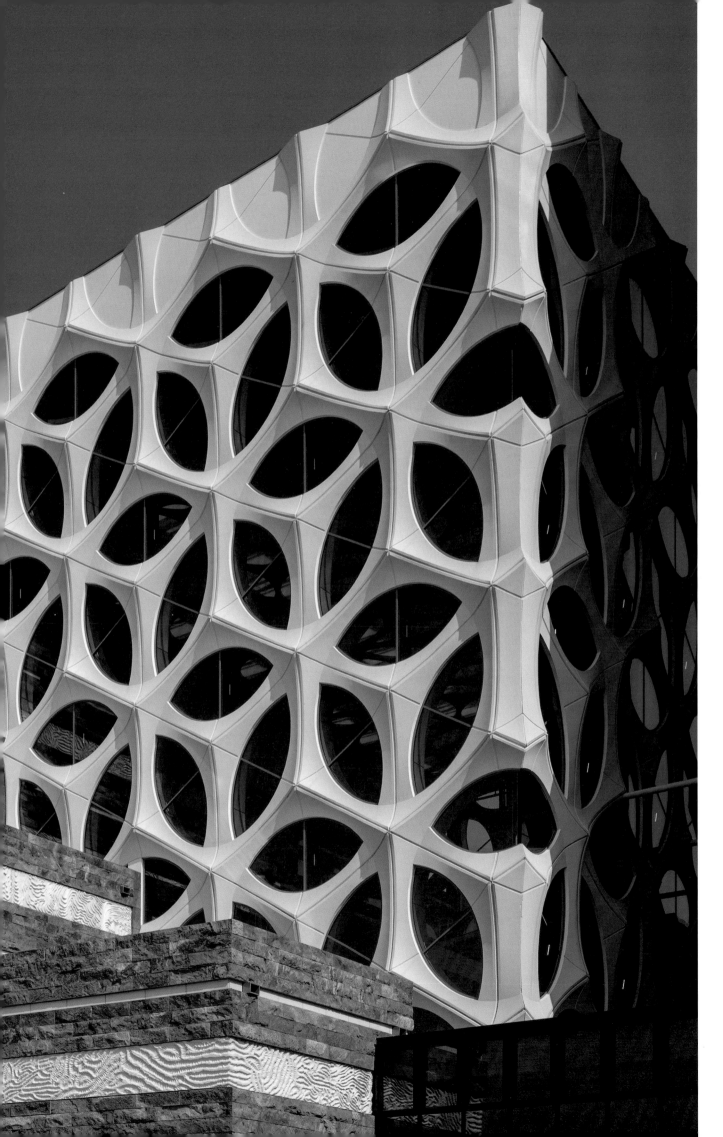

of the institute together: the existing offices and depots and the newly built museum and laboratories. The design of the atrium consists of a three-dimensional concrete structure in the form of interlocking molecules as a lace of ovals, triangles and hexagons. The filtered light that enters through the circular windows resembles a "glass crown" where scientists, staff, students and families meet, reinforcing the monumentality of the space.

Public functions such as the restaurant, the shop and the exhibition hall can be found on the ground floor where passers-by can catch sight of biologists examining whales that have washed ashore. The main staircase leading up to exhibitions resembles a mountain path, becoming narrower at the top with just enough space to welcome Trix, the 66 million-year-old T-Rex, who has been given pride of place in the Dino Era gallery.

The exterior of the exhibition halls, of stone blocks in horizontal layers, mimics a geological structure. The variety of stone used, creats a beautiful sparkle. The layers of stones are interrupted by white concrete friezes designed by famous Dutch fashion designer Iris van Herpen. Invited by Neutelings Riedijk Architects, she designed a total of 263 panels, inspired by the natural shapes of the collection. These appear to be smooth as silk, thanks to a special technique developed for Naturalis: the resemblance to fabric is a nod to the innovative dresses designed by Van Herpen for celebrities like Cate Blanchett, Beyoncé and Lady Gaga.

Inside the museum, Dutch designer Tord Boontje, known for his lighting, furniture and fabrics with exquisite floral and animal motifs, displays almost 100 striking and colorful wall panels. These are visual stories that blend photography and drawing to reveal the wonders of the natural world.

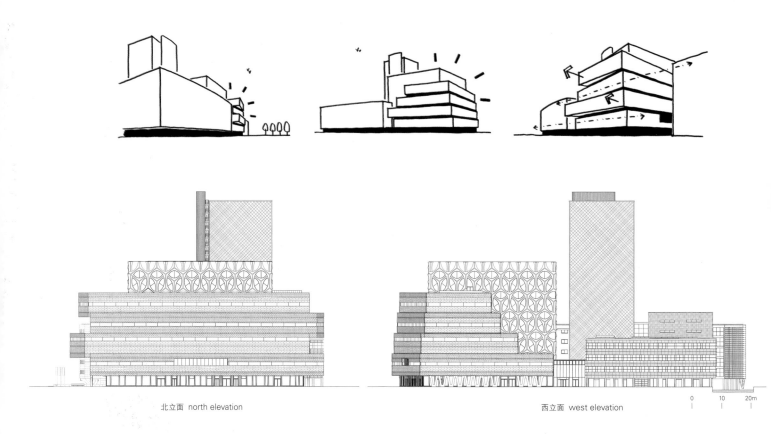

北立面 north elevation   西立面 west elevation

项目名称：Naturalis Biodiversity Center / 地点：Darwinweg 2, Leiden, The Netherlands / 事务所：Neutelings Riedijk Architecten / 设计团队：Michiel Riedijk, Willem Jan Neutelings, Frank Beelen, Kenny Tang, Guillem Colomer Fontanet, Jolien Van Bever, Inés Escauriaza Otazua, Marie Brabcová, Cynthia Deckers / 建筑工程师：ABT BV Ingenieursbureau 结构工程师：Aronsohn Raadgevende Ingenieurs / 设施设计：Huisman en van Muijen / 建筑物理：DGMR Raadgevende Ingenieurs / 总承包商：J.P. van Eesteren / 设施承包商：IC ULC-Kuijpers / 室内设计师：general public areas – Neutelings Riedijk Architecten; offices – Hollandse Nieuwe / 艺术品设计师：Iris van Herpen (betonreliëf), Studio Tord Boontje (grafiek) / 城市规划师：Studio Hartzema / 成本计算：IGG / Bointon de Groot / 景观建筑师：H+N+S, Amersfoort / 客户：Naturalis Biodiversity Center / 功能：museum 17,000m², offices en depots 18,000m², laboratories 3,000m² / 建筑面积：total of 38,000m² of which 20,000m² new built and 18,000m² renovation of existing buildings / 设计开始时间：2013.3 / 施工时间：2017.1—2019.5 / 使用时间：2019.8 / 摄影师：©scagliolabrakkee (courtesy of the architect)

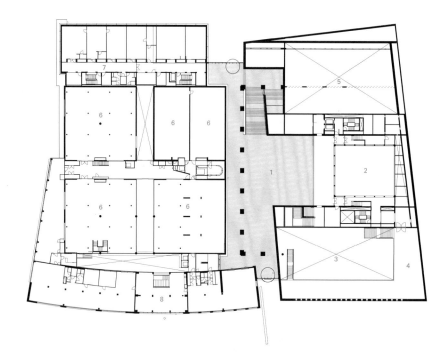

1. 中庭
2. 礼堂上空空间
3. 展厅上空空间
4. 展厅
5. 餐厅上空空间
6. 藏品存放处
7. 实验室
8. 办公室

1. atrium
2. void auditorium
3. void exhibition
4. exhibition
5. void restaurant
6. collection depots
7. laboratories
8. offices

二层 first floor

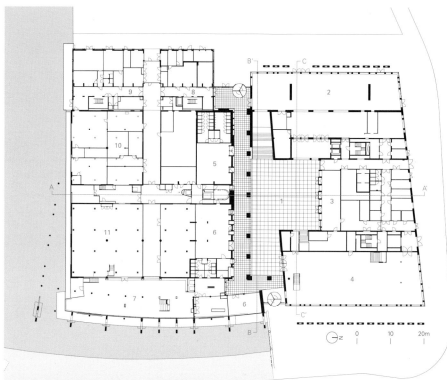

1. 中庭
2. 餐厅
3. 商店
4. 展厅
5. 衣帽间
6. 教育空间
7. 入口办公室
8. 入口实验室
9. 实验室
10. 车间
11. 书籍存放处

1. atrium
2. restaurant
3. shop
4. exhibition
5. cloakroom
6. education room
7. entrance offices
8. entrance laboratories
9. laboratories
10. workshops
11. book depot

一层 ground floor

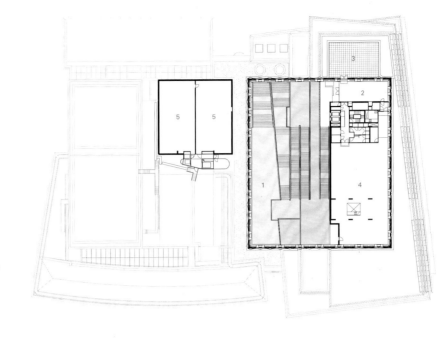

1. 中庭
2. 接待大厅
3. 屋顶露台
4. 技术设备室
5. 藏品存放处

1. atrium
2. reception hall
3. roof terrace
4. technical equipment room
5. collection depots

十层 ninth floor

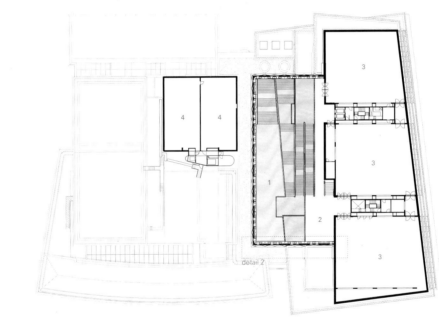

1. 中庭
2. 大厅
3. 展厅
4. 藏品存放处

1. atrium
2. lobby
3. exhibition
4. collection depots

八层 seventh floor

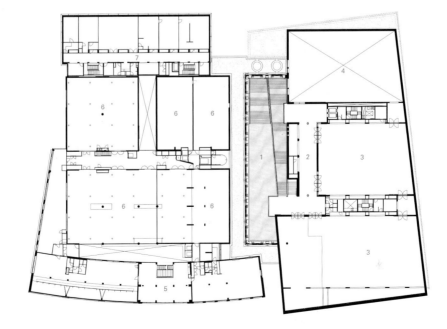

1. 中庭
2. 大厅
3. 展厅
4. 展厅上空空间
5. 办公室
6. 藏品存放处
7. 实验室

1. atrium
2. lobby
3. exhibition
4. void exhibition
5. offices
6. collection depots
7. laboratories

四层 third floor

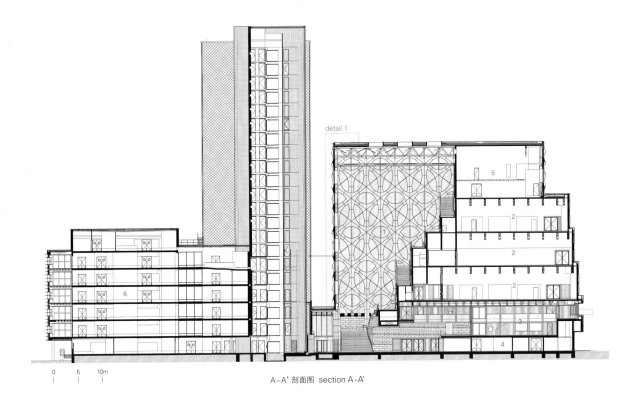

A-A' 剖面图 section A-A'

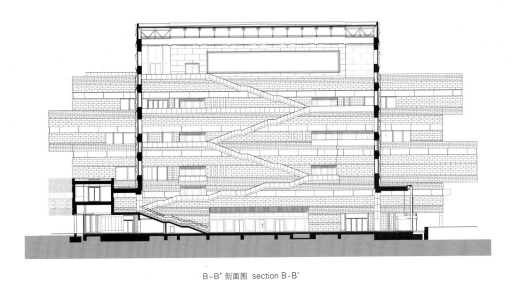

1. 中庭
2. 展厅
3. 礼堂
4. 商店
5. 技术设备室
6. 藏品存放处
7. 餐厅
8. 接待大厅

1. atrium
2. exhibition
3. auditorium
4. shop
5. technical equipment room
6. collection depots
7. restaurant
8. reception hall

B-B' 剖面图 section B-B'

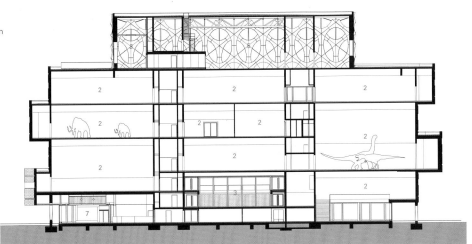

C-C' 剖面图 section C-C'

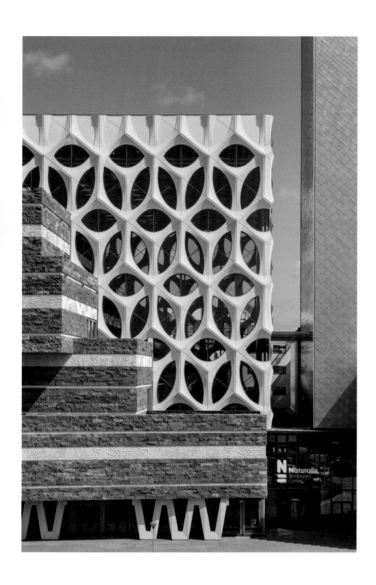
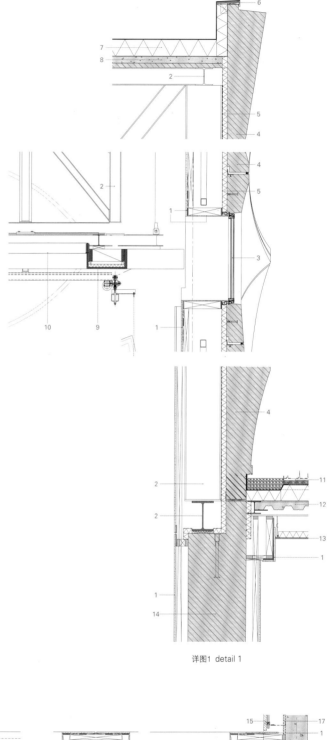
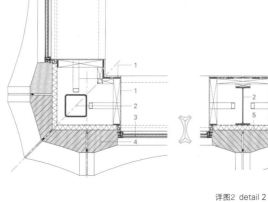
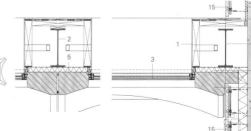

1. wall cladding: panels with oak veneer, partially perforated for acoustical absorption
2. steel structure
3. insulation glass in aluminum window frame system
4. prefab concrete panel
5. thermal insulation
6. aluminum roof edge profile
7. roofing on thermal insulation
8. concrete roof slab
9. atrium ceiling: wooden panels with oak veneer
10. double layer of translucent stretch ceiling
11. green sedum roof on thermal insulation
12. steel-concrete roof slab
13. wooden ceiling with acoustical joints
14. structural column: prefab concrete
15. natural stone panels on stainless steel anchor system
16. structural wall: in-situ concrete
17. acoustical insulation

详图1 detail 1

详图2 detail 2

AGORA 癌症研究中心
AGORA Cancer Research Center

Behnisch Architekten

Behnisch 建筑事务所设计的位于洛桑的癌症研究中心拥有雕塑般的遮阳立面
Behnisch Architekten's oncology facility in Lausanne boasts a sculptural, solar-shading facade

AGORA癌症研究中心加建的大楼位于洛桑市沃州大学医院院内，目光所及就是日内瓦湖。该中心大楼专门为400名科学家和从业人员提供了工作和研究空间。

该大楼的设计强调全面性和整体性，为癌症研究中心的研究人员提供一个便于交流以及从事癌症方面研究的场所。由于项目场地中的景观与周边景观视觉上的联系对于整体建筑体块的设计和规划起到了相当关键的作用，如同雕塑般的建筑体块与周围环境相比较就会显得辨识度很高。

由于其位置独特，人们从该市的许多地方都可以看到这一建筑物。这里又是沃州大学医院院内一处宁静的所在。它位于洛桑老城的上方，从远处看像是一座雕塑。反过来，宽敞的门厅玻璃窗可以从内部向外看到广阔的城市和全景视野的山脉。

将多种需求融入复杂且紧凑的城市肌理中对建筑和功能设计概念要求很高。建筑师将建筑的一端与周围已有的建筑联系到一起，而另一端自然地过渡到周围的自然景观中，使得建筑本身的造型不会破坏场地上原有的功能体块关系。公共区域和实验区域都有较高的空间质量，这从自然光照、建筑比例以及材料使用上都能直观地感受到。跨学科和学科之间的交流是研究能够成功的关键，这在楼层平面的组织中也能明显看出来。

建筑师们试图在该项目中实现高度复杂、呈几何形状的固定的遮阳网格设计，目标是创建一个遮阳系统，提供望向窗外时无障碍的视野，同时避免夏季阳光直射，将环境光反射到房间的深处，并使冬季的阳光可以进入室内。由于整栋建筑的形状非常不规则——几乎没有立面是垂直的，并且一个立面内也会发生水平和垂直变化，因此必须针对每种情况创建不同的构件。

AGORA癌症研究中心是受邀参加的设计竞赛的获胜项目。在设计竞赛期间，已经有足够的时间和资金来解决上述复杂的设计理念。通常，如果一个人在竞赛中努力创新，一般只能提出一些粗略的想法，使得评委们难以把握其想法的复杂性，更别说相信这些想法可以付诸实施了。网格设计遍及整座建筑。由于建筑的几何形状及其拐角是根据其他标准设计的，因此建筑师必须确保以建筑和技术上可接受的方式对网格的角落进行连接。每个构件的一部分都带有激光切割孔，以使对比不那么鲜明，从而避免太阳眩光。

The newly constructed building of AGORA Cancer Research Center is located in Lausanne on the campus of the CHUV (Center Hospitalier Universitaire) Vaudois within sight of Lake Geneva. The facility specifically provides working and research space for 400 scientists and practitioners under one roof. The major focus was on a comprehensive, holistic concept to design communication and working spaces for the cancer research center. Existing visual connections on the site and to the landscape beyond were crucial for a cautious formulation and adaption of the building mass. Thus, the building's sculptural character makes it easily recognizable within its surroundings and contributes to its autonomous presence.
Due to its location, the building can be perceived from many locations in town. Within the CHUV it embodies a place of tranquility. Situated above the old town of Lausanne, it

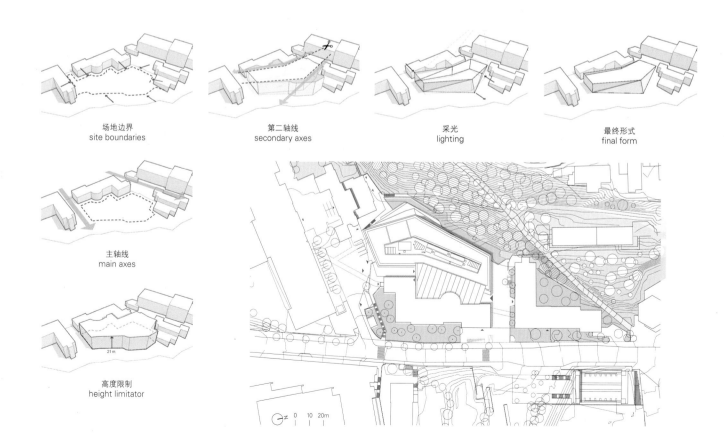

场地边界 site boundaries　第二轴线 secondary axes　采光 lighting　最终形式 final form

主轴线 main axes

高度限制 height limitator

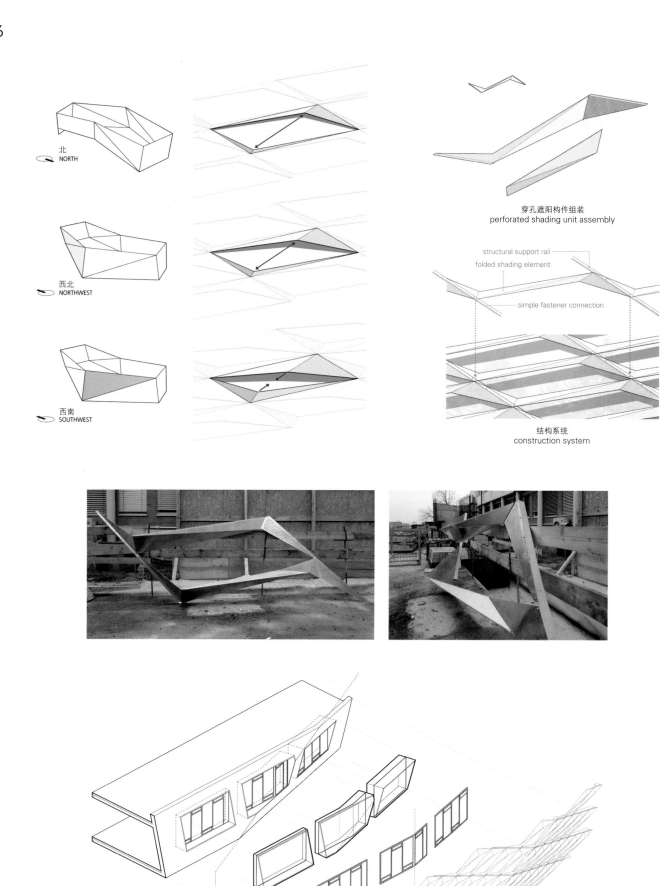

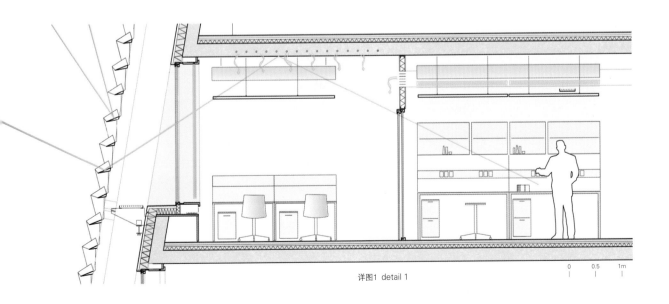

详图1 detail 1

1. 主入口 2. 中庭 3. 研讨室
1. main entrance 2. atrium 3. seminar
A-A' 剖面图 section A-A'

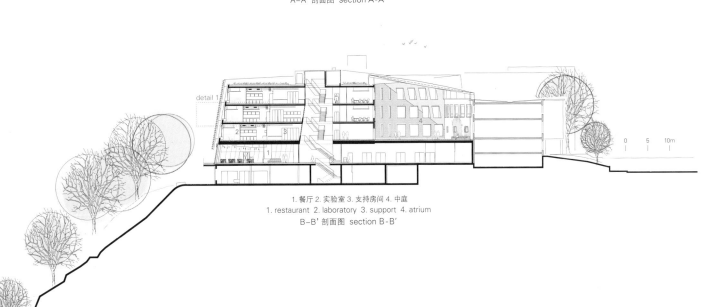

1. 餐厅 2. 实验室 3. 支持房间 4. 中庭
1. restaurant 2. laboratory 3. support 4. atrium
B-B' 剖面图 section B-B'

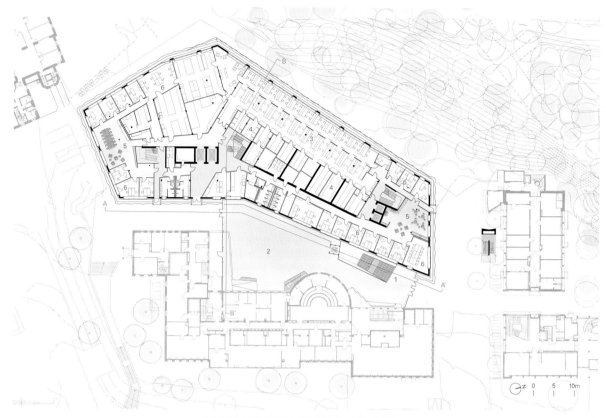

1. 主入口 2. 中庭 3. 实验室 4. 支持房间 5. 员工广场 6. 办公室
1. main entrance 2. atrium 3. laboratory 4. support 5. faculty agora 6. office
三层 second floor

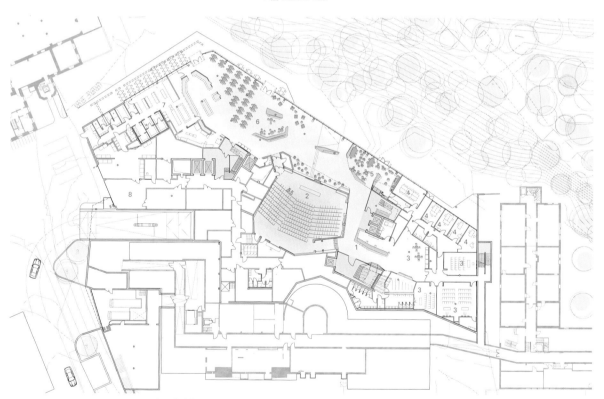

1. 主入口与接待处 2. 礼堂 3. 研讨室 4. 行政管理办公室 5. Lennards咖啡厅 6. 餐厅 7. 厨房 8. 后勤部门
1. main entrance and reception 2. auditorium 3. seminar 4. administration 5. Cafe Lennards 6. restaurant 7. kitchen 8. back of house
二层 first floor

项目名称：AGORA Cancer Research Center / 地点：Rue de Bugnon 25A, CH-1010 Lausanne, Switzerland / 事务所：Behnisch Architekten – Stefan Behnisch, Stefan Rappold / 客户：Fondation ISREC / 合伙人：Stefan Behnisch / 项目负责人：Cornelia Wust / 项目团队：Natasa Bogojevic, Ioana Fagarasan, Michael Innerarity, Matthias Jäger, Heinrich Lipp (Wbw), François Servera, Saori Yamane / 投标、施工经理：Fehlmann Architectes SA
项目管理：Cougar Management / 景观：Oxalis Architectes Paysagistes Associés / 结构工程师：ZPF Ingenieure AG / 交通：Planungsbüro Stadtverkehr, H. u. B. Schönfuß Gbr / 立面：Emmer Pfenninger Partner AG / 机电管道：AZ Ingénieurs Lausanne SA

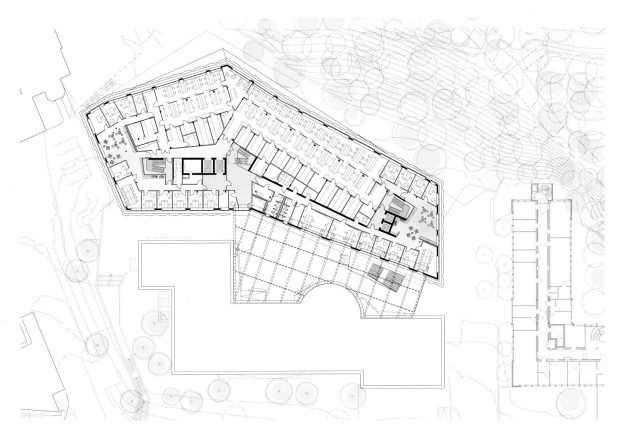

五层 fourth floor

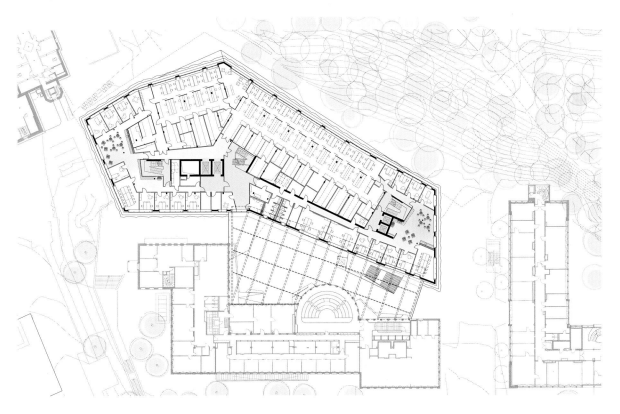

四层 third floor

实验室规划：Dr. Heinekamp Laborund Institutsplanung / 气候工程：Transsolar Energietechnik GmbH / 电气：Bering AG / 照明：Bartenbach LichtLabor GmbH
厨房技术：Bureau de planification Yvan Tercier / 声学设计：AAB J.Stryjenski & H. Monti / 消防：Swissi AG / 土工技术：Geotechnisches Institut
总承包商：Steiner SA / 建筑面积：15,700m² / 总楼面面积：22,500m² / 体积：93,000m³ / 工作位数量：400 /
停车位数量：183 / 建筑规模：three stories below ground, four stories above ground / 设计竞赛时间：2012 / 规划开始时间：2013 / 施工时间：2015—2018
摄影师：©David Matthiessen (courtesy of the architect), except as noted

emerges as a sculptural element when seen from afar. In turn, the generous glazing of the foyer allows a wide view outward from the interior that encompasses both the city and the panorama of the mountain scenery.

Connecting the diverse requirements into an extremely heterogeneous and tight urban fabric places great demands on the architectural and programmatic concept. The decision to link to an existing building on one side while creating a harmonious transition to nature on the other, informs the building's structure without adversely affecting the site's existing programmatic functions. Spatial qualities, directly perceived through daylight, proportion and materiality, are equally visible in public areas and in the highly technical laboratories. Interdisciplinary as well as disciplinary communication are both central to successful research and are evident in the organization of the floor plans.

The architects tried to implement the idea of a highly sophisticated, geometrical, fixed sun-shading grid in this project. The goal was to create a sun protection system which allows almost unobstructed views to the outside and – at the same

time – keeps the summer sun out, reflects the ambient light into the depth of the room, and allows the winter sun partially to enter the space. Since AGORA's building mass is very freely-shaped – hardly any facade is vertical, and horizontal and vertical changes occur within one facade – a different element had to be designed and created for each situation.

The AGORA Cancer Research Center was acquired through an invited competition. During the competition there was already sufficient time and funds to solve this sophisticated concept. Often if one strives to be innovative in competitions, typically one can only present rather sketchy ideas, making it difficult for jurors to grasp their complexity, let alone trust that these ideas can actually be implemented. The mesh extends all around the building. Since the geometry of the building and thus its corners were designed according to other criteria, the architects had to make sure that the grid was optimized in a way that its corners meet in an architecturally and technically acceptable way. A part of each element bears laser-cut holes in order to minimize contrast and thereby avoid the solar glare.

CSN 总部办公楼扩建项目
Expansion of the CSN Headquarters

BGLA + NEUF Architectes

CSN 将总部的规模扩大了一倍，所有员工可以在同一屋檐下工作
CSN expands the size of its head office double to bring all the employees together under one roof

建筑如何为民主价值观做出贡献？

这个LEED银奖扩建项目新近获得了Grands Prix du Design公共空间类设计大奖，为CSN总部提供了一种全新的生活，提升了使用者和周边居民的生活质量。

CSN总部位于Ville-Marie区，Jacques-Cartier桥的东部，是一座20世纪80年代早期的野兽派建筑。建筑师希望去除它沉重严肃的建筑形象，重新开发和扩建，改善其建筑形象。建筑师在保留其原有本质和特征的同时增强了建筑的美感和对周边社区的影响。整个项目的灵感来源于CSN一直以来的价值观：自主、自由、团结。

概念方法

扩建项目的目标是提升这一区域的建筑与社交景观。一个彩色的巨大体量成为原有建筑的延续，但相对更加自由、表现多样、具有互补性和统一性。

新建的花园完善了场地空间，为使用者和周边居民提供了一个真正放松和交流的区域。

该建筑核心设计理念是两个体量的共生，木材体量与钢材体量通过一个沐浴在自然光线的大中庭和谐地融合在一起。这是一个自由、明亮、统一的地方，是一个真正的公共空间，起到了强有力的连接作

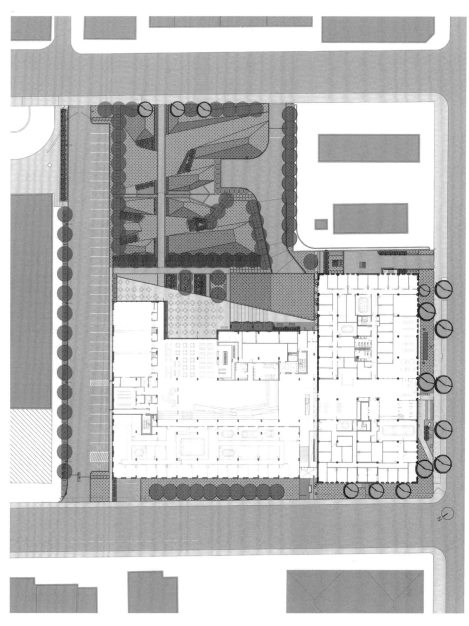

立面+立面研究
elevations + facade study

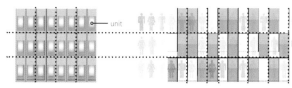

原有 existing   扩建 expansion

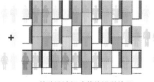

单独设计组成整体视觉效果
individual contributions to a collective vision

用,既使扩建结构与原有建筑连接,又让联盟成员之间建立联系。办公室都位于扩建体量的外侧,走廊和连接桥沿着中庭布置。一层和花园层是公共功能区,包括会议室、自助餐厅、多功能室、书店和图书馆。"圆环"平面具有很大的灵活性,能够在长时间的使用中,根据联盟成员的增长或减少进行调整。

增加木结构的使用

木材、混凝土和钢材展现出这一地方和区域特有的工业,也代表了CSN工会。中庭的设计目的也在于展示这些原材料。特别需要注意的是,木材是这个项目最主要的元素,是建筑公共空间的建筑材料。这些空间对联邦来说具有高度象征意义,它代表着不同的成员之间通过一个共同的使命团结凝聚在一起。精心使用的木材在共享空间中得到强调,如中庭、自助餐厅、多功能室。采用精致工艺的木材与混凝土、钢材相结合,增加了集体空间的美感,代表了工作者丰富的专业知识,呼应了CSN的本质。为项目挑选的不同类型木材都来自以联邦工会为代表的地方和区域工业。枫树由于其颜色、强度和特性合适,是项目内部广泛使用的装饰材料。

How can architecture contribute to democratic values?

Recently awarded at the Grands Prix du Design, category "common area", this LEED silver expansion project offers a new everyday life at the CSN (Confédération des syndicats nationaux)'s head office, thus improving the quality of life of users and residents of the neighborhood.

Located in the Ville-Marie district, east of the Jacques-Cartier Bridge, the head office of the CSN, a brutalist building from the early 80s, wanted to get rid of its heavy and severe image. This image has been transformed through a redevelopment and an expansion amplifying the aesthetics of the building and its impact on the neighborhood, while preserving its identity. The project is inspired and anchored in the continuity of the values of the CSN: autonomy, freedom, solidarity.

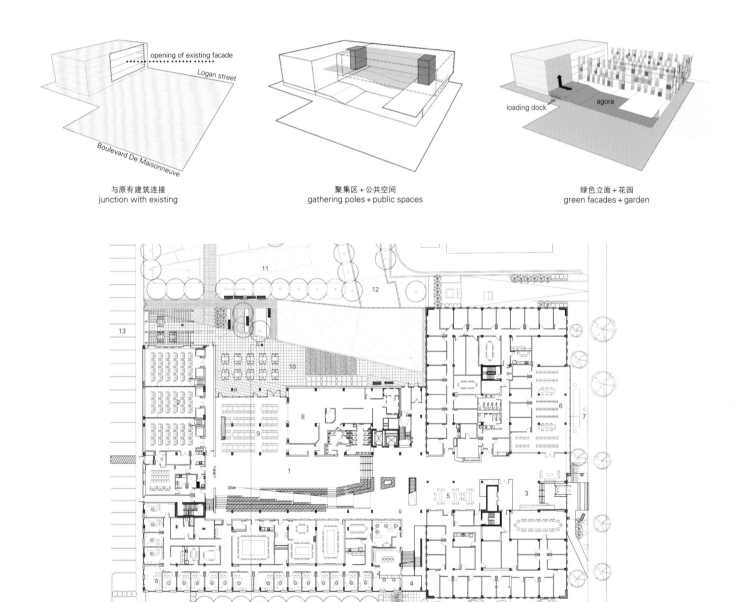

与原有建筑连接
junction with existing

聚集区 + 公共空间
gathering poles + public spaces

绿色立面 + 花园
green facades + garden

1. 中庭 2. 聚集空间 3. 主入口（Lormier街）4. 次入口（Logan街）5. 接待区 6. 图书馆/书店 7. 展示窗 8. 自助餐厅 9. 餐厅 10. 广场 11. 花园 12. 装载区 13. 停车场
1. atrium 2. gathering spaces/multi-functional room 3. main entrance (Lormier street) 4. secondary entrance (Logan street) 5. reception area 6. library/bookstore 7. display window 8. cafeteria 9. dining room 10. agora 11. garden 12. loading dock 13 parking lot

一层 ground floor

项目名称：Expansion of the CSN Headquarters / 地点：1601 de Lorimier Ave., Montréal, Québec, Canada, H2K 1M5 / 事务所：BGLA + NEUF Architectes
景观建筑：Projet paysage / 其他专家与顾问：mechanical, electric engineering – Ingémels Experts-Conseils; structural engineering – Pasquin St-Jean; Building commissioning – TST; contractor – EBC Inc.; building code – Technorm; laboratory, environment – Sanexen; laboratory, geotechnical – InspecSol; building envelope – Édifice Expert; elevator consultant – Exim
客户：Confédération des syndicats nationaux (CSN) / 总楼面面积：20,545m² / 建筑造价：39,5 M$ / 竣工时间：2017.7 / 摄影师：©Stéphane Brügger (courtesy of the architect)

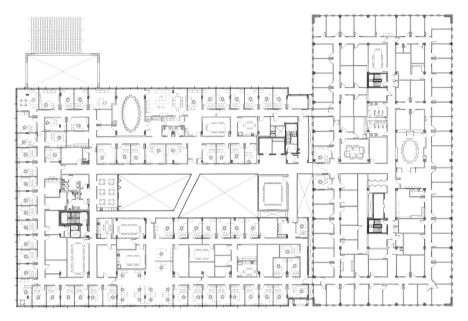
四层 third floor

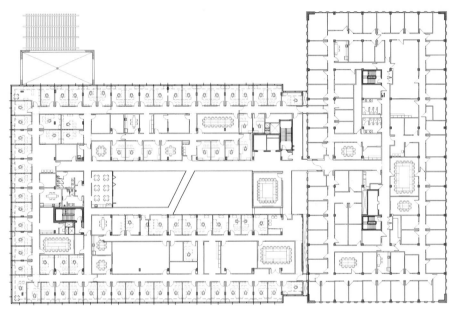
三层 second floor

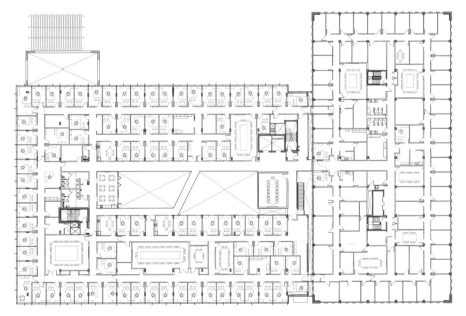
二层 first floor

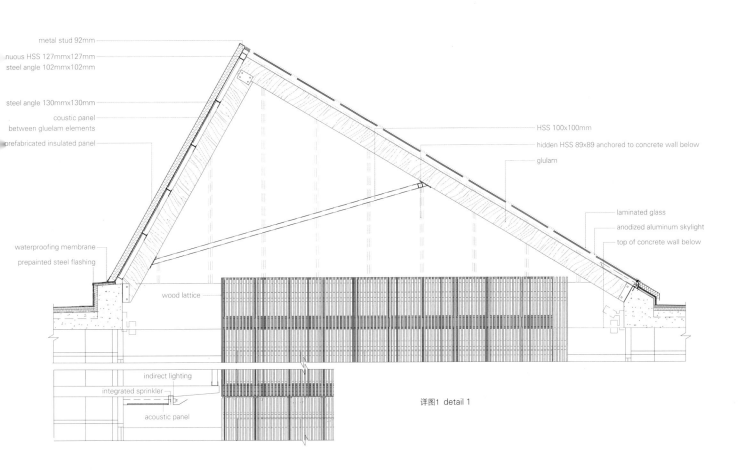

详图1 detail 1

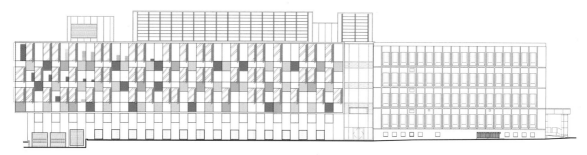

西立面 west elevation

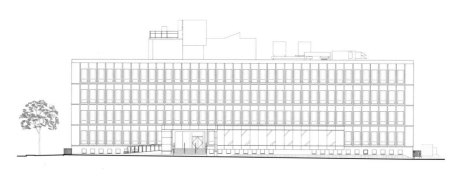

南立面 south elevation

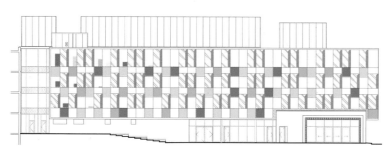

东立面 east elevation

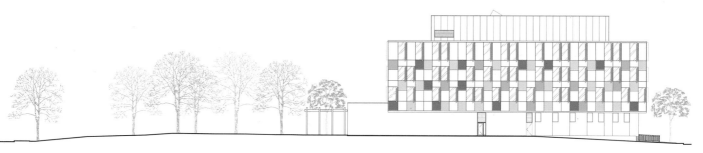

北立面 north elevation

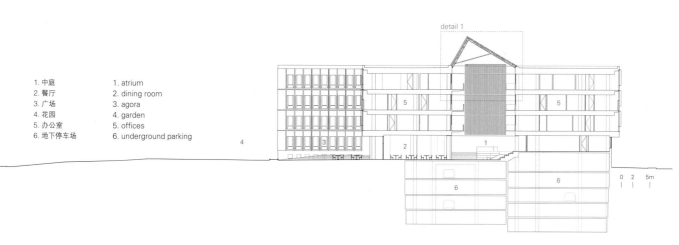

1. 中庭　　　　1. atrium
2. 餐厅　　　　2. dining room
3. 广场　　　　3. agora
4. 花园　　　　4. garden
5. 办公室　　　5. offices
6. 地下停车场　6. underground parking

A-A' 剖面图 section A-A'

Conceptual Approach

The expansion project is designed to enhance the architectural and social landscape of the area. A large volume of color develops in continuity with the existing building, but now more freely, expressing diversity, complementarity and unity. A new garden completes the development of the site and offers users and residents of the neighborhood a real area of relaxation and exchange.

At the heart of the architectural concept is the symbiosis of two volumes, one of wood, the other of steel, blending harmoniously in a vast atrium bathed in natural light. A true metaphor of the public space, at once free, luminous and unifying, this grandiose place acts as a strong link, uniting the existing building with its expansion, as well as the different members of the confederation. The offices are all located at the perimeter of the enlarged volume and around this atrium lined with passageways and crossed bridges. On the ground floor and on the garden level are the common functions: the meeting rooms, the cafeteria, the multifunctional room, the bookshop and the library. The "circular" floor plan offers great flexibility that will allow the growth or decline of the different federations over the years.

Enhancing Wood

Wood, concrete and steel are used to express the know-how of local and regional industries, represented by CSN unions. The atrium is designed to highlight these raw materials. In particular, wood is the predominant element of this project, since it is associated with the collective spaces of the building; these places, highly symbolic for the Confederation, unite its different members and gather them around the same mission. The careful use of wood allows emphasis to be placed on these spaces of sharing, including the atrium, the cafeteria and multifunctional rooms. Combined with concrete and steel, finely worked wood enhances the aesthetics of these collective spaces and represents the wealth of know-how of the workers, recalling the essence of the CSN. The different types of wood chosen for the design of the project all come from local and regional industries represented by the Confederation's trade unions. Maple is the main type of wood used in various ornamental applications for its color, strength and identity.

日内瓦鲁道夫·斯坦纳学校
Rudolf Steiner School of Geneva

LOCALARCHITECTURE

鲁道夫·斯坦纳学校的扩建项目重新发现了让·雅克·屈米原始设计的精髓
Rudolf Steiner School extension rediscovers the essence of Jean-Jacques Tschumi's original designs

崛起的精神

考虑如何扩建学校的出发点是学校本身:对学校的场地、通行路线和建筑本身等的质量和缺陷进行分析研究,但也要考虑到这个地方的精神、灵感和愿望。现在的学校建筑由建筑师让·雅克·屈米在20世纪80年代后期设计建造。其特点是一系列有机统一的建筑围合成一个庭院,所有的交通流线空间都围绕庭院而流动。朝内的露天庭院只是朝外的教学空间的序曲。教学空间面向外面景观,遵循太阳一天中的运行轨迹而设计布局。

一个顶点

处理这样一个项目——给建筑增加高度,并富有活力地体现该建筑的交通流线、人们的运动方式以及使用它的人,是一个特别的挑战。首先,它涉及对学校精神的理解和尊重,并寻求重新发现最初设计的本质,围绕一个中心点进行根本的开发设计。

要与原设计师屈米的设计有机地结合在一起,给建筑师提出了一个难题。建筑师通过以下步骤来解决这一难题:从顶部楼梯的平台处修了2级台阶,通到延伸至半空中的墙壁处,院子里的模型柱廊也朝着天空延伸,所有这些都预示着还应有延续部分存在——即一个顶点,缺失的连接部分可能就在场地与天空之间。

七间新教室

该设计延长了主楼梯,通向包含七间新教室的新楼层。原有的混凝土结构支撑着新框架。新建的有顶过道由框架以及进入教室的门围合而成。教学空间由3D合成的9面教室平面图和新设计的开有大型天窗的屋顶构成,教室立面拥有大型玻璃窗,可以俯瞰Salève山。教室全部镶着木板。

体验之旅

当你漫步在这栋建筑中,各种建筑构件的设计组合方式就显而易见了。身临其境,你可以感受到空间的有机特性,建筑的每个构件都强调了学校在其环境中的连贯一致性。因此,建筑与环境的关系不是静态的、一成不变的,而是不断发展的、灵活的、自然的,是以目的为导向的,在不同层次上体现其适应性和内在智能。

新愿景核心的工艺

新结构几乎完全由瑞士木材制成。3D技术的使用融合了先进的建造方法和古老的专业知识,创造出一种新的协作模式和一种超越了传统做法但与此共享的建筑语言,从而创建了创新的、鼓舞人心的解决方案。数字设计和生产技术已经将三十年前图纸上所勾画的设想变为了现实。

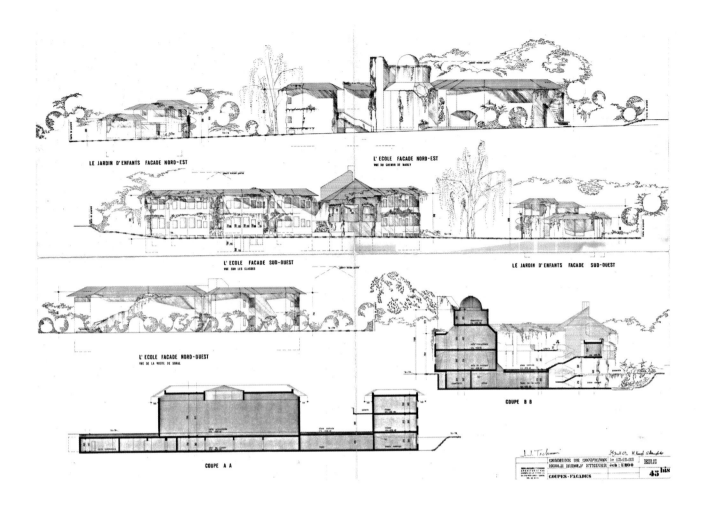

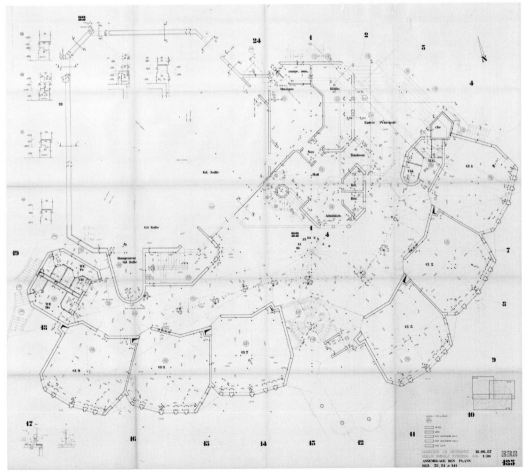

让·雅克·屈米画的原图，1983—1987年 original drawings by Jean-Jacques Tschumi, 1983~87

一所鲜活的学校

这个全新的一整个楼层的教室扩建项目是在学校继续像往常一样持续运行的情况下实现的,项目工作时间表与学校假期和学校生活的节奏同步。新楼层的设计还考虑了周围的景观——周围的小区环境与最初的建筑建成时相比,已经发生了变化。

Rising Spirit

The starting point for considering how to expand the school was the school itself: an analytical examination of the site, its access routes and buildings, their qualities and defects – but also a taking account of the spirit of the place, its inspirations and aspirations. The school buildings, designed and realized by the architect Jean-Jacques Tschumi in the late 1980s, are characterized by their organic forms, enclosing a courtyard towards which the circulation spaces flow. The inward-looking, uncovered, courtyard offers a prelude to the outward-looking teaching spaces, oriented towards the landscape and following the sun's path through the sky.

A Culmination

Tackling a project like this – adding height to a building epitomized by its vibrant expression of circulations, movements and the people using and embodying it – posed a particular challenge. First of all, it involved understanding and respecting the spirit of the school, and seeking to rediscover the essence of the initial design, with its radical development around a central point.

Working organically with Tschumi's design presented the architects with a conundrum: two steps on a staircase rising from the final landing and ending in a wall suspended in mid-air, and the colonnade of modeled pillars on the courtyard extending upwards towards the sky, all seemed to anticipate a continuation – a culmination. The missing link, perhaps,

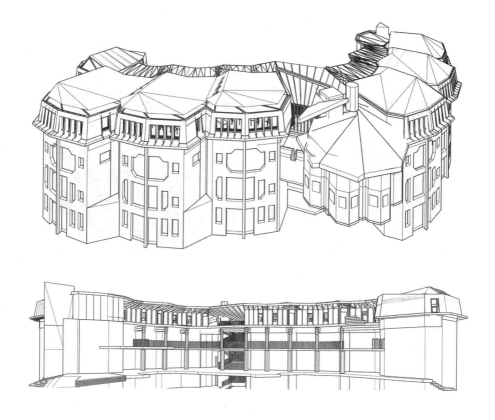

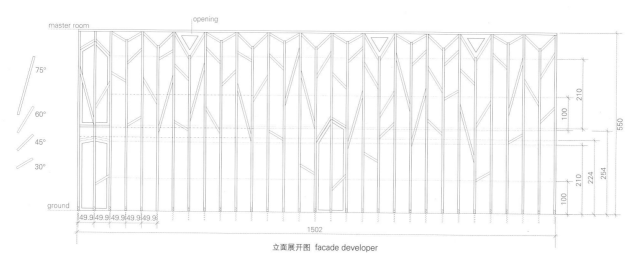

立面展开图 facade developer

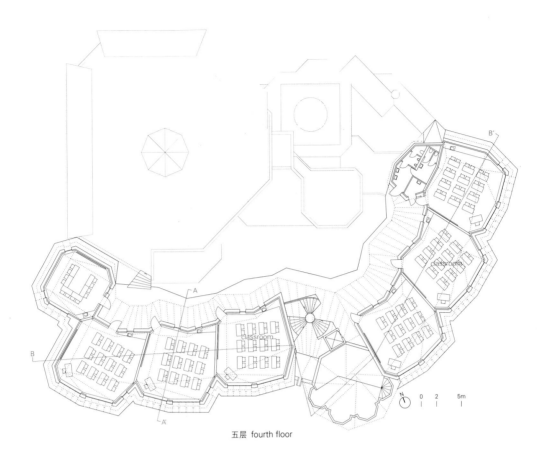

五层 fourth floor

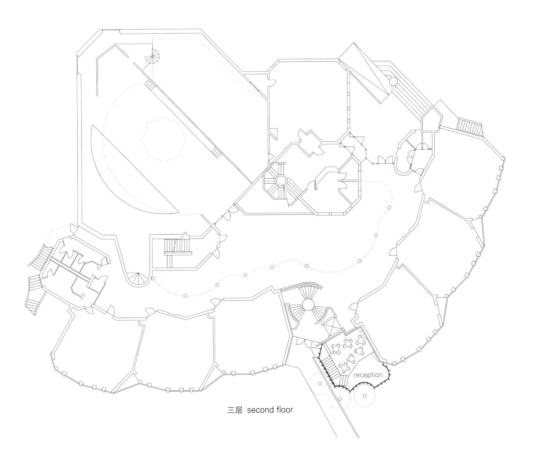

三层 second floor

between the site and the sky above it.

Seven New Classrooms
The main staircase is extended, leading to a new story comprising seven classrooms. The existing concrete structure supports the new framework. The new covered walkway is structured by the consoles of the framework and the doors to the classrooms. The teaching spaces represent a 3D synthesis of the nine-sided classroom floorplans and the new roof incorporating large skylight windows, while a large glazed facade overlooks the Salève mountain. The classrooms are entirely paneled in wood.

An Experiential Journey
The way that the various elements are assembled to form a structure becomes apparent when moving within it. The organic nature of the space is revealed in a journey, each element of the building underscoring the school's coherence within its environment. The building's relation to the environment is therefore not a fixed and static phenomenon but something evolving, flexible, natural and attentive to purpose, attributing qualities of adaptability and intrinsic intelligence to its different layers.

Craft at the Heart of a New Vision
The new structure is made almost entirely of Swiss wood. The use of 3D technology brought together cutting-edge methods and age-old expertise, creating a new mode of collaboration and a shared language which moved beyond conventional practices to create innovative, inspirational solutions. Digital design and production technologies have fulfilled aspirations expressed on paper thirty years ago.

A Living School
This construction of a completely new floor of classrooms was realized as the school continued to operate as usual, the schedule of works arranged to synchronize with holidays and the rhythms of school life. The design of the new floor also took account of the surrounding landscape – a neighborhood that had changed since the initial construction project.

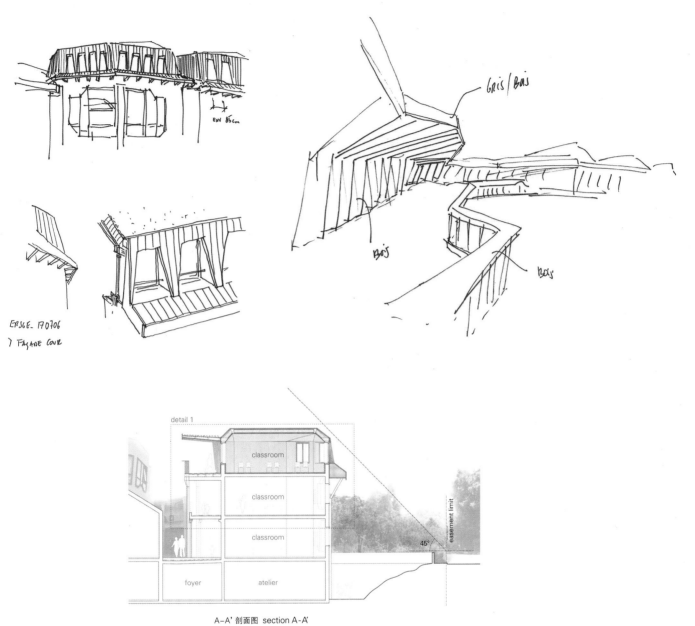

A-A' 剖面图 section A-A'

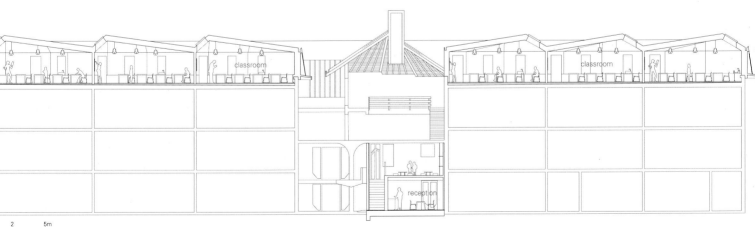

B-B' 剖面图 section B-B'

项目名称：Rudolf Steiner School of Geneva / 地点：Chemin de Narly 2, Confignon (GE), Switzerland / 事务所：Localarchitecture
项目团队：Antoine Robert-Grandpierre, Manuel Bieler, Laurent Saurer, Pedro Vieira, Francesca Aiello, Tim Cousin / 施工管理：Thinka Architecture
土木工程师：INGENI SA / Heating pump engineer: SIG / 暖通卫生电气与安全工程师：srg.eng Carpenter: Ateliers Casaï /
客户：École Rudolf Steiner de Genève / 建筑面积：650m² / 体积：3,363m³ / 概念设计与实现时间：2015—2018 / 摄影师：©Matthieu Gafsou

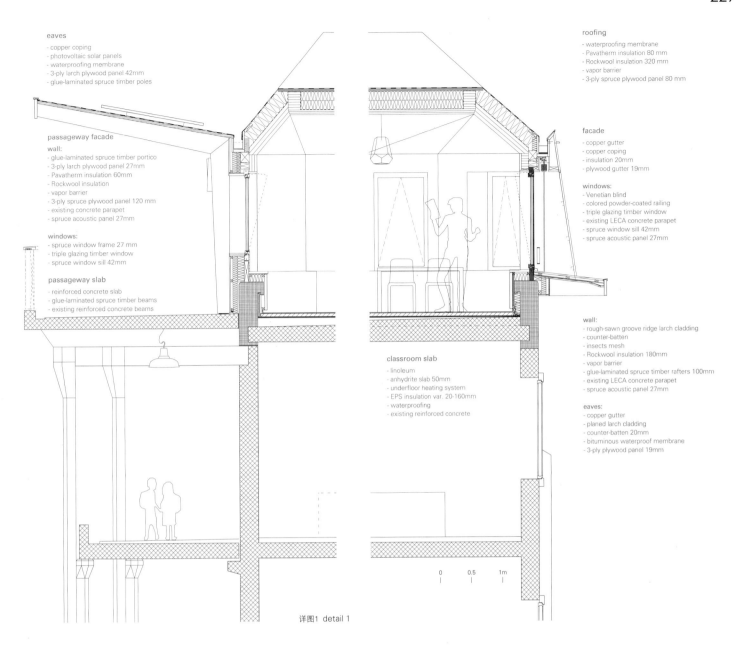

eaves
- copper coping
- photovoltaic solar panels
- waterproofing membrane
- 3-ply larch plywood panel 42mm
- glue-laminated spruce timber poles

passageway facade
wall:
- glue-laminated spruce timber portico
- 3-ply larch plywood panel 27mm
- Pavatherm insulation 60mm
- Rockwool insulation
- vapor barrier
- 3-ply spruce plywood panel 120 mm
- existing concrete parapet
- spruce acoustic panel 27mm

windows:
- spruce window frame 27 mm
- triple glazing timber window
- spruce window sill 42mm

passageway slab
- reinforced concrete slab
- glue-laminated spruce timber beams
- existing reinforced concrete beams

classroom slab
- linoleum
- anhydrite slab 50mm
- underfloor heating system
- EPS insulation var. 20-160mm
- waterproofing
- existing reinforced concrete

roofing
- waterproofing membrane
- Pavatherm insulation 80 mm
- Rockwool insulation 320 mm
- vapor barrier
- 3-ply spruce plywood panel 80 mm

facade
- copper gutter
- copper coping
- insulation 20mm
- plywood gutter 19mm

windows:
- Venetian blind
- colored powder-coated railing
- triple glazing timber window
- existing LECA concrete parapet
- spruce window sill 42mm
- spruce acoustic panel 27mm

wall:
- rough-sawn groove ridge larch cladding
- counter-batten
- insects mesh
- Rockwool insulation 180mm
- vapor barrier
- glue-laminated spruce timber rafters 100mm
- existing LECA concrete parapet
- spruce acoustic panel 27mm

eaves:
- copper gutter
- planed larch cladding
- counter-batten 20mm
- bituminous waterproof membrane
- 3-ply plywood panel 19mm

详图1 detail 1

P74 Paulo Mendes da Rocha

Was born in 1928 in Vitória, Brazil. Whole family moved to São Paulo in 1940. Graduated from the Mackenzie Architecture School in 1954. After graduation, he won the competition in 1957 for the construction of a gymnasium, the "Clube Atlético Paulistano". This work brought him public recognition and won the "Grande Prêmio Presidência da República" award at the 6th São Paulo Biennial in 1961. In 1968, he won the competition for the Brazilian Pavilion at Osaka Expo 70. Has been awarded with the Mies Van der Rohe Foundation Prize in 2000, the Pritzker Architecture Prize in 2006 and the Golden Lion for lifetime achievement at 2016 Venice Architecture Biennale.

Lacol

Is an architecture cooperative set up in 2014, Barcelona. Currently, is a team of 14 people who work in various fields of architecture. The main axis of work on which they apply the knowledge of other fields is the cooperative housing. The most relevant case has been 'La Borda' housing cooperative, where, in addition to carrying out the architectural project, they have participated in its promotion and all the learning process has been systematized through the book *Habitar en comunidad* (2018). They are currently building 'La Balma', a similar housing cooperative in Poblenou, and accompanying several projects in the promotion phase in other parts of Catalonia. In order to promote and make viable the replicability of the model, they work in the design of public policies, advise municipalities, and write articles in different publications.

P206 BGLA

BGLA has been specializing for more than 40 years in the fields of architecture, heritage and urban design. Now has three business offices in Montreal, Quebec City, and Sept-Îles, with a large team of more than 70 employees. Has a diversified experience in the institutional (teaching and health), cultural and community fields. Its architectural approach is sensitive to the protection of the built environment and its historical and social context. Is concerned with the valorization of the natural and cultural heritage of each place.

LOCALARCHITECTURE

Was founded by Manuel Bieler[right], Antoine Robert-Grandpierre[left] and Laurent Saurer[center] in Lausanne, Switzerland in 2002. Manuel Bieler received his M. Arch in 1996 and attended a Postgraduate course of Construction Economy and Managment at the EPFL. Antoine Robert-Grandpierre also received a M. Arch in 1996 at the EPFL. Is President of the Maison de l'Architecture, Geneva since 2013. Laurent Saurer has worked at the Geninasca & Delefortrie Architects, MPH Architects and Workshop 15X10. Is a Member of the Commission of the Federation of Swiss Architects since 2013. They are Member of the Federation of Swiss Architects (FSA) since 2010. Recent awards include the International Wood Architecture Award in 2019 and Best Architects Award in 2018 and 2015. They has been involved in teaching for several years, in particular as lecturer.

P74 **MMBB Arquitetos**
Was founded in 1991 as a result of the association of the architects Fernando de Mello Franco, Marta Moreira, and Milton Braga. Currently partered with Marta, Milton and Maria João Figueiredo, consolidating a comprehensive professional performance, which has stood out by the development of public and institutional designs in the area of building, infrastructure and urbanism. Led by the concern with a critical and analytical posture, extends their activities to the cultural and academic sphere. Not only carry out an associate parallel production in the way of participation and/or organization of cultural events, exhibitions, biennials but also develop an academic activity encompassing teaching and research. MMBB won several distinctions for its projects; the 9th Ibero-American Architecture and Urbanism Biennale in Rosario (2014), the São Paulo Association of Art Critics Awards (2013), the 3rd International Architecture Biennale Rotterdam (2007), the International Architecture Biennale of São Paulo, and the Institute of Architects of Brazil.

P206 **NEUF Architectes**
Founded in 1971, NEUF is one of the largest architecture and design firms in Canada. Specializes in finding creative solutions to today's most challenging design problems with offices in Montreal, Ottawa and Toronto. Has completed the CBC Headquarters in Ottawa, Bombardier facilities in Mirabel and Dorval. Received multiple design awards recently at the World Architecture Festival in Berlin and Amsterdam.

P110 **Smiljan Radic**
Was born in Santiago, Chile in 1965. Graduated in architecture from Universidad Católica de Chile and studied at the Istitutto Universitario di Architettura di Venezia(IUAV), Italy. Travelled for three years and finally he opened his own studio in 1995 in Chile. In 2001, he was named 'Best Architect Under 35' by the Chilean Architects Association, and selected as a part of Architectural Record Design Vanguard 2008. Has been a lecturer and honorary member of the American Institute of Architects since 2009.

P110 **Eduardo Castillo**
Was born in Chile, 1972. Graduated from Catholic University of Chile in 2000 and after then started his career as an architect. Received a Doctorate in Architecture and Urban Studies in 2005. Is currently a professor of Architecture at his alma mater. In 2011 he won the first prize in the Biobío Regional Theater competition with Smiljan Radic and Gabriela Medrano.

P110 **Gabriela Medrano**
Was born in Ecuador, 1982. Graduated from Andrés Bello National University in Santiago de Chile with bachelors and masters degrees in Architecture. Received a degree in sustainable architecture from Catholic University of Chile in 2009. Won the first prize in the Biobío Regional Theater competition with Smiljan Radic and Eduardo Castillo in 2011. Was selected as the first prize winner in the Santiago Communication Tower competition with Smiljan Radic and Ricardo Serpell in 2014. Is Professor in the School of Architecture at the Universidad San Sebastian from 2017. Has worked with Smiljan Radic since 2008.

P128 **Mecanoo architecten**
Officially founded in Delft in 1984, is led by creative director/founding partner, Francine Houben[picture-above], design and research director/partner Dick van Gameren, and technical director Friso van der Steen. The extensive collective experience results in designs that are realised with technical expertise and great attention to detail. Discovers unexpected solutions based on process, consultation, context, urban scale, and integrated sustainable design strategies, the practice creates culturally significant buildings with a human touch. Mecanoo's projects range from single houses to complete neighbourhoods, cities and polders, schools, theatres and libraries, hotels, museums, and even a chapel.

P92 **BIG**
Founded in 2005 by Bjarke Ingels, BIG is a Copenhagen, New York and London based group of architects, designers, urbanists, landscape professionals, interior and product designers, researchers, and inventors. Currently gets involved in a large number of projects throughout Europe, North America, Asia, and the Middle East. Believes that in order to deal with today's challenges, architecture can profitably move into a field that has been largely unexplored. A pragmatic utopian architecture that steers clear of the petrifying pragmatism of boring boxes and the naïve utopian ideas of digital formalism. Like a form of programmatic alchemist, it creates architecture by mixing conventional ingredients such as living, leisure, working, parking, and shopping. By hitting the fertile overlap between pragmatic and utopia, once again finds the freedom to change the surface of our planet, to better fit contemporary life forms.

Jean Nouvel was born in Fumel, France in 1945. Obtained his degree at ENSBA (Ecole Nationale Supérieure des Beaux-Arts), Paris in 1972. Started his first architecture practice in 1970. His works have gained world-wide recognition through numerous prestigious French and International prizes and rewards. In 1989, The Arab World Institute in Paris was awarded the Aga-Khan Prize. He received the Lion d'Or of the Venice Biennale in 2000 and the Royal Gold Medal of the RIBA, the Praemium Imperial of Japan's Fine Arts Association and the Borromini Prize given to architects under 40 in 2001. Was appointed Docteur Honoris Causa of the Royal College of Art in London in 2002. Also received the International High-rise Award 2006 and prestigious Pritzker Prize in 2008. In France, he received many prizes including the Gold Medal of the French Academy of Architecture, two Équerre d'Argent and National Grand Prize for Architecture.

Is Dean of IE School of Architecture and Design. Since 2005, she has been serving as the Executive Director of the Pritzker Architecture Prize. Prior to joining IE University, she was Associate Curator of the department of architecture at The Art Institute of Chicago. Is the co-author of the books *Masterpieces of Chicago Architecture* and *Skyscrapers: The New Millennium*, and author of numerous articles for architectural journals and encyclopedias. Received a master of city planning degree from the University of Pennsylvania and a bachelor of arts degree in urban affairs from the State University of New York at Buffalo. Currently serves on an international jury for the award, 'ArcVision: Women and Architecture', a prize honoring outstanding women architects.

P30 Sou Fujimoto Architects

Sou Fujimoto graduated from the Department of Architecture, Faculty of Engineering at the Tokyo University. He established Sou Fujimoto Architects in 2000. His recent work, Mille Arbres won the competition of Reinventer Paris with Manal Rachdi OXO in 2016. He also won the 1st prize in the New Learning Center at Paris-Saclay's Ecole Polytechnique in 2015. His representative works are 'the Serpentine Gallery Pavilion 2013', 'House NA (2011)', 'Musashino Art University Museum & Library (2010)' and so on.

Manal Rachdi, Nicolas Laisne, Sou Fujimoto, Marie de France, and Dimitri Roussel, from left.

P30 OXO Architectes

Manal Rachdi founded OXO Architectes to develop architectural systems transforming the research, behavior analysis and theoretical thought into conceptual driving forces. He collaborated with well-known architects such as: Duncan Lewis (2003-2005), the Agence Du Besset-Lyon, to finally joining the Ateliers Jean Nouvel in 2007. He participated and led the competitions for: the Seoul Opera, the Philarmonie of Paris in 2008, and the 53W53 competition for the mix used program tower for the MOMA in New York.

P148 Phil Roberts

Is a design writer based in Montreal, Canada. Also works as a design consultant for various companies in the creative industries. Has an Honours Bachelor of Arts from the University of Toronto, where he majored in architectural design and minored in Canadian studies and Spanish.

P56 Angelos Psilopoulos

Studied architecture at the School of Architecture, Aristotle University of Thessaloniki (AUTh), then moved on to his Post-Graduate studies at the National Technical University in Athens (NTUA). Is currently pursuing his Ph.D. at the NTUA on the subject of Theory of Architecture, studying gesture as a mechanism of meaning in architecture. Has been working as a freelance architect since 1998, undertaking a variety of projects both on his own and in collaboration with various firms and architectural practices in Greece. Since 2003, he has been teaching Interior Architecture and Design in the Department of Interior Design, Decoration, and Industrial Design at the Technological Educational Institute of Athens (TEI).

P8 Anna Roos

Studied architecture at the UCT (University of Cape Town) and holds a postgraduate degree from the Bartlett School of Architecture, UCL, London. Moving to Switzerland in 2000, she worked as an architect, designing buildings in Switzerland, South Africa, Australia, and Scotland. As a freelance architectural journalist since 2007, besides *C3*, she also writes for *A10*, *Ensuite Kultur Magazin*, *Monocle*, and *Swisspearl* architecture magazine. Her first book, *Swiss Sensibility: The Culture of Architecture in Switzerland* (2017), recently published by Birkhauser Verlag, is also available in German and French editions.

Stefan Behnisch ©Christoph Soeder

Was founded in 1989 and works out of three offices – Stuttgart, Munich, and Boston. These offices are directed by Stefan Behnisch and his partners Robert Hösle (Munich), Robert Matthew Noblett (Boston), Stefan Rappold and Jörg Usinger (Stuttgart). Has a global reputation for high-quality architecture that integrates environmental responsibility, creativity, and public purpose. The firm produces a rich variety of buildings mainly in Europe and North America. The five partners and staff share a vision to push the boundaries of high performance, 21st-century architecture that respects user needs, ecological resources, and local cultures.

P174 **Neutelings Riedijk Architects**

Dutch architect and professor, Michiel Riedijk (1964) received his Master in Architecture at the TU Delft in 1989. After graduation, he has been the director at Riedijk Bekkering Architecten until 1991. In 1992, he established the Neutelings Riedijk Architects in Rotterdam with Willem Jan Neutelings. He has been enrolled in the Dutch Architects Register as architect since 1993. Since 2007, he is also Professor of Public Building at the Faculty of Architecture and the Built Environment at his alma mater. Some exemplar works are the MAS Museum in Antwerp, the Netherlands Institute for Sound and Vision in Hilversum, and the Culture House 'Rozet' in Arnhem.

P62 **Studio Hollenstein**

Is an architecture and urban design studio founded in Sydney in 2010 and is led by Matthias Hollenstein. Was rebranded in 2019 from Stewart Hollenstein to Studio Hollenstein, to reflect its renewed direction under Matthias Hollenstein. Enjoys working across a range of project types and scales in Australia and internationally. Is driven by a desire to create strategic, innovative and joyful places that respond to the challenges and opportunities of user's time and invite them to inhabit them. Received the AR Library Award, The Chicago Atheneum Architecture Award, The AIA NSW Premier's prize, and multiple AIA awards for architecture, urban design and interior architecture.

P154 **NADAAA**

Is a Boston-based architecture and urban design firm led by principal designer Nader Tehrani, in collaboration with partner Katherine Faulkner. After Post-graduate Program at AA Graduate School of History and Theory, Nader Tehrani received his M. Arch in Urban Design with Distinction from GSD Harvard. Served as Head of the Department of Architecture, MIT from 2010-2014. Is the Dean of the Irwin S. Chanin School of Architecture at the Cooper Union in New York since 2015. Katherine Faulkner received her M. Arch from GSD Harvard, and an MBA from Boston University. Recently received the 2019 AIA COTE Top 10 Award and the 2017 Women in Design Award of Excellence. Is a member of AIA, Boston Society of Architects, and Urban Land Institute.

Matthias Hollenstein | Nader Tehrani ©Leo Sorel | Katherine Faulkner courtesy the Holcim Foundation

© 2020 大连理工大学出版社

版权所有·侵权必究

图书在版编目(CIP)数据

公共建筑与空间营造 / 丹麦BIG建筑事务所等编；于风军等译. — 大连：大连理工大学出版社，2020.12
ISBN 978-7-5685-2690-6

Ⅰ．①公… Ⅱ．①丹… ②于… Ⅲ．①公共建筑—空间设计 Ⅳ．①TU242

中国版本图书馆CIP数据核字(2020)第173025号

出版发行：大连理工大学出版社
　　　　　（地址：大连市软件园路80号　邮编：116023）
印　　刷：上海锦良印刷厂有限公司
幅面尺寸：225mm×300mm
印　　张：14.75
出版时间：2020年12月第1版
印刷时间：2020年12月第1次印刷
出 版 人：金英伟
统　　筹：房　磊
责任编辑：杨　丹
封面设计：王志峰
责任校对：张昕焱
书　　号：978-7-5685-2690-6
定　　价：298.00元

发　行：0411-84708842
传　真：0411-84701466
E-mail：12282980@qq.com
URL：http://dutp.dlut.edu.cn

本书如有印装质量问题，请与我社发行部联系更换。

墙体设计
ISBN: 978-7-5611-6353-5
定价：150.00元

新公共空间与私人住宅
ISBN: 978-7-5611-6354-2
定价：150.00元

住宅设计
ISBN: 978-7-5611-6352-8
定价：150.00元

文化与公共建筑
ISBN: 978-7-5611-6746-5
定价：160.00元

城市扩建的四种手法
ISBN: 978-7-5611-6776-2
定价：180.00元

复杂性与装饰风格的回归
ISBN: 978-7-5611-6828-8
定价：180.00元

内在丰富性建筑
ISBN: 978-7-5611-7444-9
定价：228.00元

建筑谱系传承
ISBN: 978-7-5611-7461-6
定价：228.00元

伴绿而生的建筑
ISBN: 978-7-5611-7548-4
定价：228.00元

微工作·微空间
ISBN: 978-7-5611-8255-0
定价：228.00元

居住的流变
ISBN: 978-7-5611-8328-1
定价：228.00元

本土现代化
ISBN: 978-7-5611-8380-9
定价：228.00元

都市与社区
ISBN: 978-7-5611-9365-5
定价：228.00元

木建筑再生
ISBN: 978-7-5611-9366-2
定价：228.00元

休闲小筑
ISBN: 978-7-5611-9452-2
定价：228.00元

景观与建筑
ISBN: 978-7-5611-9884-1
定价：228.00元

地域文脉与大学建筑
ISBN: 978-7-5611-9885-8
定价：228.00元

办公室景观
ISBN: 978-7-5685-0134-7
定价：228.00元